THE BEE-KEEPER'S I

CW01498828

OR

TWENTY-TWO YEARS' EXPERIENCE

IN

QUEEN-REARING.

CONTAINING THE ONLY SCIENTIFIC AND PRACTICAL METHOD OF REARING
QUEEN BEES, AND THE LATEST AND BEST METHODS FOR THE
GENERAL MANAGEMENT OF THE APIARY.

BY HENRY ALLEY.

PRACTICAL APIARIST.

ILLUSTRATED.

PUBLISHED BY THE AUTHOR.
WENHAM, MASS.
1883.

PRINTED AT THE SALEM PRESS,

cor. Liberty and Derby Sts.,

SALEM, MASS.

This book is due on the date indicated
below and is subject to an overdue fine
as posted at the Circulation Desk.

Harold A. Sholl

1-10-40

PREFACE.

It may not be out of place for me to give the reasons that induced me to assume the position of author and give to the bee-keeping fraternity a work of this kind. I had been engaged in queen-rearing but a few years when I became convinced that the system then in vogue was faulty and that a much better method was needed and might be devised. With this object in view I at once instituted a series of experiments which, I am happy to say, have been crowned with success, even beyond my most sanguine expectations, and I am now able to present to my fellow bee-keepers a system which is both scientific and practical.

After so many years of arduous toil and study in perfecting this method, I felt that it was no more than just that I be reasonably remunerated for my labors and for the knowledge that I had gleaned from my long experience in queen-rearing; consequently I decided to publish my method in book-form, trusting to the hearty response of the bee-keeping fraternity for remuneration.

It will be remembered that the Rev. L. L. Langstroth (a bee-master whose name we love to honor)

148327

had intended to publish a similar work, but poor
health and the loss of his son obliged him to abandon
it, much to the regret of his many friends and ad-
mirers.

I need not state that this work was not intended
as a literary effort, as indeed I make no pretensions
in this respect. I have endeavored to present to my
readers a work that will be beneficial and advantageous to them and have tried to avoid all that is
superfluous or ambiguous, believing that plain, practical common sense is far preferable; and if it meets
with general approval I shall rest content. I claim
that my method of rearing queens is new and original, being the results of my long experience in
queen-rearing and practical apiculture.

In order that my readers might have the benefit
of the experience of one of the most prominent and
successful apiarists in the country, I obtained from
Mr. George W. House, of Fayetteville, N. Y., an
essay on the best method of managing the apiary in
order to obtain the largest amount of surplus honey,
including preparing for market and marketing the
same. It is ably written and instructive, presenting
many new and original ideas which are sound, practical and invaluable, and which cannot fail to benefit either the novice or the expert, if the directions

given are carefully put into practice. Mr. House seldom fails to secure a fine crop of honey and is successful in marketing the same.

I have also included an essay on the new races of bees from the pen of Silas M. Locke, of Salem, Mass., who, during the season of 1881, was employed by Mr. D. A. Jones, of Beeton, Ont., as superintendent of the queen-rearing department. Mr. Locke has had the best of opportunities to study the markings, habits and characteristics of the new races, and his essay will be found interesting and instructive, presenting many new features regarding them, and forming one of the best descriptions yet given to the public.

By the careful study of this work, and by put-, into practice the directions herein given, one will experience no difficulty in rearing queens of a superior quality. Let it ever be our aim to rear *better* not *cheaper* queens.

AUTHOR.

CONTENTS.

(vii)

LIST OF ILLUSTRATIONS.

INTRODUCTION.

THE IMPORTANCE OF QUEEN—REARING AS A BRANCH OF APICULTURE.

TWENTY-FIVE years ago I purchased my first colony of bees, and with that event began my interest in apiculture. My colony, being in a box hive, I transferred to frames, and commenced to rear queens and to experiment generally with bees. In the course of a few years, upon the introduction of the Italian bee to this country, there came a large and increasing demand for bees of this race.

Many bee-keepers began to rear queens and to offer bees and supplies for the apiary for sale, a large number of whom soon failed for want of patronage, or were compelled to abandon the business on account of the cheap queen traffic. Of all who as late as ten years ago were engaged in this branch of the business, I can call to mind but one beside myself. Few have made queen-rearing and the supply trade a success.

Shortly after the introduction of the Italian bee, the "American Bee Journal" sprang into existence, and simultaneous with its appearance began one of the most important industries of the age, viz.: practical apiculture. At this early stage of its history, queen-rearing was in its infancy, while but few bee-keepers had any practical knowledge of this interesting and

1 (1)

vastly important branch of the business, and apparently very little advancement has been made up to the present time as compared with the other branches of apiculture.

After a thorough examination of the latest works on the subject, and a careful study of all the various Bee Journals, I find only the old methods taught which were in use many years ago. Hence the bee-keeping public continue to rear queens in the old way, the result being that a vast number of inferior and even worthless queens are put upon the market every season.

The present and future interests of apiculture demand a more thorough and practical method of rearing queens, and I shall endeavor in this work to give my readers such information as shall tend to give a new impetus to this branch of bee-keeping, and also aid, if possible, in doing away with the cheap and worthless queens produced under the lamp nursery system; and to offer to the bee-keeping public, for their careful consideration and adoption, a thorough, practical and scientific method of queen-rearing, which is the result of many long years of practical experience, and much hard study.

In order to become a successful instructor one must first attain a complete knowledge of the subject to be taught, and unless it has been thoroughly and fully mastered in all its details, failures only can result.

In presenting this work to the bee-keeping fraternity, I do not wish to assume the position of teacher, but rather to place before its readers in as plain and practical a manner as possible my method of rearing queens, leaving to their judgment the careful study, and candid criticism of its contents, feeling assured of a favorable decision regarding its merits and value; knowing that if its instructions are carefully studied in all their details, and put to a practical test, the result will be successful. By careful attention to all the rules laid down herein, I hope better queens will be produced, a matter of great importance to the bee-keeper whether he keeps bees for pleasure or profit; and of vastly more importance to the bee-master who follows it as a vocation and depends upon the same for a living.

CHAPTER I.

THE SELECTION AND CARE OF BREEDING QUEENS.

All bee-masters know the importance of having a strong, vigorous and healthy queen for the mother bee; hence it will be admitted that all queens used for this purpose should be carefully selected, perfect in every respect, and of undoubted purity and prolific-ness. The particular strains from which we wish to breed should be thoroughly tested, to determine their qualities regarding purity, gentleness, honey-gathering, and wintering. This is very important and essential as some do not combine all the very desirable requisites above mentioned, and I would lay it down as a rule to breed only from such.

THE ADVANTAGE OF USING SMALL HIVES AND COMBS FOR THE BREEDING QUEENS.

If a large number of queens are to be reared, the mother queens should not be kept in full colonies, as the risk of killing them in securing eggs for cell-building is too great, and many valuable queens are lost in this way. To guard against such loss my breeding queens are kept in miniature hives (fig. 1). The queen being more easily and speedily found on small combs, there is much less risk of injuring her, and there will be eggs enough in one of them at any time to start fifty or more cells.

FIG. 1. Fertilizing or miniature hive.

COMBS TO USE IN OBTAINING EGGS.

In selecting the combs for the queen to lay in, to be used expressly for cell-building, take only such as are nearly new or that have been made use of for brood but once or twice. I do not use the combs in standard frames for this purpose, as in the course of the season a large number of nice brood-combs would be either badly mutilated or destroyed in so doing. Small pieces of comb the size of the nucleus frame, described in another place, are generally at hand and far preferable to larger combs. One standard Langstroth frame of comb will fill four or five of such small nucleus frames.

FOUNDATION FOR CELL-BUILDING.

It would be a good plan to fill frames with foundation, have it worked out in full colonies, and used for brood once or twice, then cut up and fastened into the small frames. I have found foundation very good for starting cells, even when it had not been used for brood at all; but care should be taken that the cells on both sides are drawn out to nearly the proper depth. These combs, when used to obtain eggs for cell-building, will be filled so that they can be removed as often as once every twenty-four hours. They should be properly numbered and dated for future use as wanted; the other four combs in the miniature hive being used for storing honey and brood to keep the colony prosperous in young bees. A good prolific queen will fill this small comb in less than twenty-four hours, but it is better to let it remain that length of time. The advantage of using such small hives will be seen at a glance, as it will not be necessary to open a full colony every time eggs are needed from which to start cells. Again, the exact age of all eggs is easily and exactly determined, and the apiarist may tell at any time just when to prepare his bees for cell-building; the time when the queens will hatch from the cells may also be determined within a few hours. This is a matter of great importance; saving as it does, much time, labor and anxiety, as well as for other reasons which any intelligent bee-keeper will readily comprehend.

THE DISADVANTAGE OF KEEPING BREEDING QUEENS IN FULL COLONIES.

It may be convenient for those who wish to rear a few queens to open a full colony for eggs, but as they can seldom be found of the right age in sufficient quantities to start a large number of cells at any one time, the former plan is much to be preferred by large breeders.

By placing a comb, selected for the purpose, in the centre of the brood-chamber of a full colony, we can sometimes find eggs which are suitable for the purpose, on the fourth day after, which should invariably be used on that day for cell-building. Sometimes the bees will not allow the queen to use such combs in which to deposit her eggs for several days, as perhaps breeding is not going on vigorously, and the queen may not reach the empty comb for one or two days after it is placed in the brood-chamber; consequently the hive would have to be opened often to ascertain this fact. Under such circumstances, queen-rearing cannot be carried on as systematically and successfully as by the miniature hive system.

MINIATURE HIVES FOR BREEDING QUEENS.

A colony in such a hive can always be depended upon for eggs at any time for cell-building. I find that a comb put in at night will be filled with eggs the next day. Then it may be removed and another inserted in its place. Once in twenty-four hours is often enough to change them. By this plan one queen will furnish eggs sufficient to rear 10,000 queens in the course of one season. The above plain and simple rules are the first steps which should be taken and put into practice systematically, if one wishes to rear first-class queens, and make the business successful and remunerative.

CHAPTER II.

ROOM FOR HANDLING BEES; WHAT COLONIES TO SELECT FOR CELL-BUILDING; HOW TO PREPARE THE BEES FOR THIS PURPOSE.

Every bee-keeper should have a room specially adapted for the purpose of handling bees, as many of the operations about the apiary cannot at all times be carried on in the open air, especially in wet and cool weather.

In fact, most of the work about preparing bees for cell-building can be performed only in a convenient and handy place. In view of this I have, in another place, given a description of such a room.

Always select the strongest colonies for cell-building, and never the weak or feeble ones, as such would not rear strong and hardy queens. You may perchance have some strong colonies in the apiary, having queens which you wish to supersede; colonies having old, uneven or crooked combs; odd-sized frames, or those in box hives. These may be used to good advantage; thus ridding the apiary of such undesirable stocks, and should be selected in preference to those

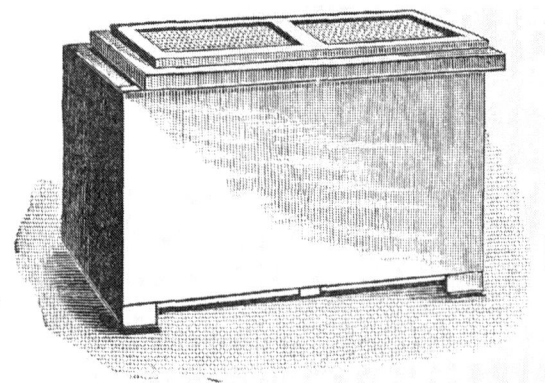

Fig. 2. Swarming box.

in good hives and in fine working condition. The combs can be transferred to other frames, and the brood given to weak colonies. Having made your selection, take them into the bee room, give the bees a few puffs of smoke to cause them to fill with honey and remove the combs, examining them carefully to find the queen; after caging her, brush all the bees into a box,—a Langstroth hive cap will answer every purpose. If the bees should attempt to crawl out blow a little tobacco smoke on them, and they will remain quiet. The bees should now be put into another box, say one that will hold three pecks, the top and bottom of which should be covered with wire cloth (fig. 2, description farther on), in order to give the bees plenty of air. The top or cover of this box may be secured by Van Dusen clamps or some other simple arrangement. I use four screws for this purpose, but seldom fasten the top on unless the box is to be transported some distance.

THE LENGTH OF TIME TO KEEP THE BEES QUEENLESS.

The bees should be kept in this box at least ten hours. Soon after being put into it they will miss their queen, and keep up an uproar until released. This prepares them for cell-building. I find it a good plan to keep them in a cool, dark room, or cellar, until needed, as they will keep more quiet and there is less danger from suffocation. If the bees were properly drummed before being removed from the hive, they have filled themselves with honey sufficient to last during the time that they are kept confined in the box; but to guard against starvation, which might occur, as the bees do not in all such instances fill themselves readily, I give them a pint of syrup in the glass feeder (fig. 8, described farther on). The bees must be kept queenless for from ten to twelve hours, else the eggs given them for cell-building will be destroyed. This I have learned from practical experience.

ROYAL JELLY; HOW PREPARED; COMPOSITION OF.

When eggs are given too soon after the bees are made queenless, they are very apt to remove or destroy them. This, however, is not the case when larvæ are provided instead. I have had some experience in keeping pigeons as well as bees, and have noticed that there is a natural food or chyle (white, milky and very nourishing) secreted in the crop of the hen, with which to feed the young and tender birds during the early stage of their development. Reason and experience teach me that the same rule applies equally well to bees, and that when made queenless and confined in the swarming box they prepare or secrete the white, milky food, which we find in the bottoms of the cells around the eggs given them for rearing queens, and which is of the same nature as the royal jelly upon which the young queens feed while confined to the cell; also that it is necessary that they be kept queenless until instinct impels them to make this important preparation for cell-building.

In support of the above I would say, first, that this secretion is found in all animal bodies (under certain circumstances) for this purpose; second, that the hen pigeon is incapable of secreting this food until about the fourteenth day of incubation, showing that the secretion is not made until needed, and lastly, the fact that bees, after being kept queenless ten or twelve hours, and then given the prepared eggs, will place this food in the bottom of the cells within an hour, going to work contentedly, knowing that they have means to produce other queens, and showing no further symptoms of queenlessness.

Again, the amount of jelly-food furnished a larva from which a queen is to be reared is much greater than that from which a worker is produced, and the composition of each is entirely different. It may not be generally known that a colony deprived of its queen almost invariably selects a larva which is usually over twenty-four hours, and frequently from two to three days old, instead of an egg from which to rear another. Such queens must necessarily be "short-lived" as they are not reared in accordance with natural laws. Otherwise, so far as known, they follow nature. Every queen cell should be so abundantly supplied with royal jelly that after the queens have hatched there will be more or less left in the cells. This is the case with the best cells produced by the bees under the swarming impulse, and I claim that just as good cells can be produced by the method which I have instituted. Quite a distinguished writer made the statement some time ago in one of the bee journals that "artificial (or forced) queens left no jelly in the bottoms of the cells." He evidently jumped at this conclusion without thoroughly testing the matter. I admit that as artificial queens are generally reared, this will be the result, but when the reader has become thoroughly conversant with the directions given in this book, and has carefully put them into practice, he can produce those that show a goodly amount of jelly in the bottom of every cell from which a queen has hatched.

CHAPTER III.

CELL-BUILDING.

Everything being now in readiness for cell-building, we go to the hive in which the breeding queen is kept and take from it the middle comb, placed there four days previous. We shall find in the bottom of the cells, if examined closely, a white, shiny substance. This is the just-hatched egg. Unless a powerful glass is used we shall be hardly able to see the small grub at this early stage of development.

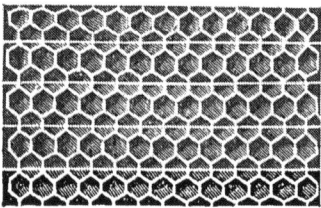

FIG. 3. Comb containing eggs.

We now take this comb into the bee-room which has been warmed to prevent the eggs from being chilled. Cut it into strips with a thin, sharp knife (an old-fashioned table knife ground thin answers the purpose), running the knife through each alternate row of

cells as seen in fig. 3.[1]   After the comb has been
cut up, lay the strips flat upon a board or table, and
cut the cells on one side down to within one-fourth
of an inch of the foundation or septum, as seen in fig.
4.[2]   A very thin, sharp and warm knife must be used,
or the cells will be badly jammed and mangled.
While engaged in this work I keep a lighted lamp
near at hand with which to heat the knife, as a delicate
operation of this kind cannot be performed well with
a cold tool, nor so quickly.

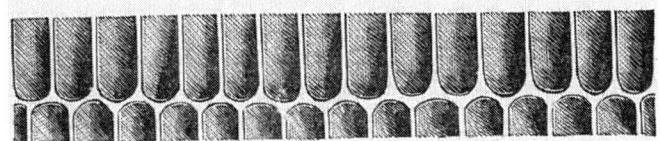

FIG. 4. Strip of comb on which cells are to be built.

HOW TO HAVE CELLS BUILT IN ROWS; THE NEW METHOD.

Now we come to a process which I have for many
years been trying to effect, and its discovery has proved
invaluable to me.   In truth, I would have paid
one hundred dollars at any time during the first sixteen
years of my experience for a knowledge of this fact.

---

[1] Fig. 3 represents a piece of comb containing eggs, with lines running
through each alternate row of cells, and showing the manner in which the
comb should be stripped up for cell-building.   This piece of comb was
also photographed, and is about one-half the natural size.

[2] Fig. 4 represents a sectional view of one of the prepared strips as
cut from fig. 3, and ready to place in position for cell-building.

I never was troubled in getting all the cells which I needed built in the old way, but to have them so evenly spaced that each could be cut out without injuring or destroying its neighboring cell, puzzled me for many years; by persistent thought, persevering labor and experiment, however, the matter was at last settled satisfactorily. This one fact alone is worth one hundred dollars to any queen-dealer, and ten times the cost of this book to any bee-keeper, even though they rear but few queens; and I feel assured that all my readers will admit this when they have tested it.

HOW TO PREPARE THE COMB FOR THE QUEEN CELLS.

The strips of comb being all ready, we simply destroy each alternate grub or egg, as seen in fig. 5.[3] In order to do this, take the strips carefully in the left hand and insert with the right the brimstone end of a

FIG. 5. Comb containing eggs in alternate cells.

common lucifer match into each alternate cell, pressing gently until it touches the bottom, and then twirl it rapidly between the thumb and finger; by this means the egg or grub will be destroyed. This gives plenty of room for large cells to be built, and the

---

[3] Fig. 5 represents the prepared strips with the egg removed from each alternate cell.

bees to work around them and also permits of their
being cut out without injury to adjoining cells, fig. 6⁴
(a full description of which is given elsewhere).

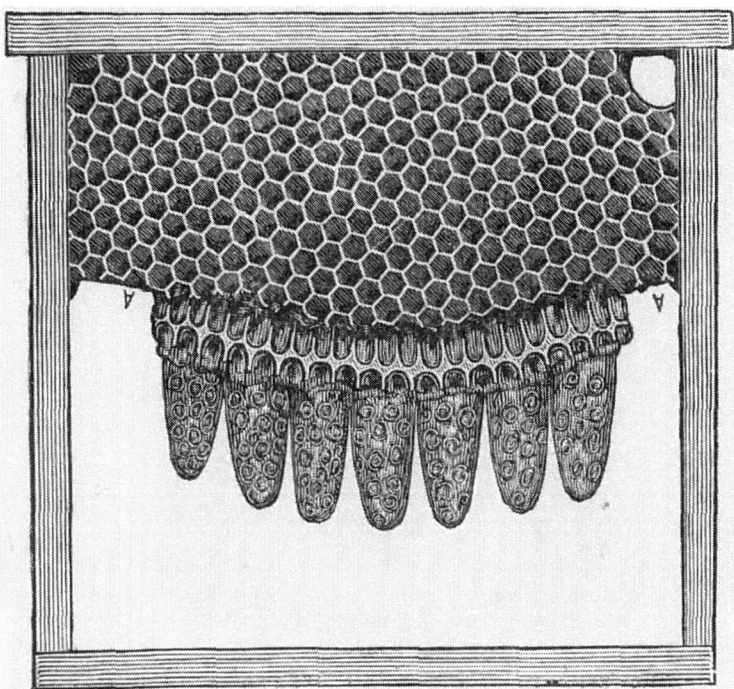

FIG. 6.   The new way of having cells built.

All queen-dealers know that this cannot be done when
the cells are built by the old method as shown in

---

⁴ This cut represents one of the frames used in my fertilizing hives,
one-half of the comb being cut out to make room for the queen-cells. A A
represents the strip of comb containing the eggs (on which the cells are
built) fastened to the comb. The frame was photographed (as shown in cut)
smaller than its natural size. The cells are exactly as the bees build them
by my new method.

fig. 7.[5] I think I hear some "doubting Thomas" say, How will you place or secure this slender strip of comb in a frame so as to prevent its falling down? I would reply, have at hand a sheet iron pan about six inches long, three inches wide, and three inches deep, being rather larger at the top than bottom (or any other sort of iron vessel that will hold hot

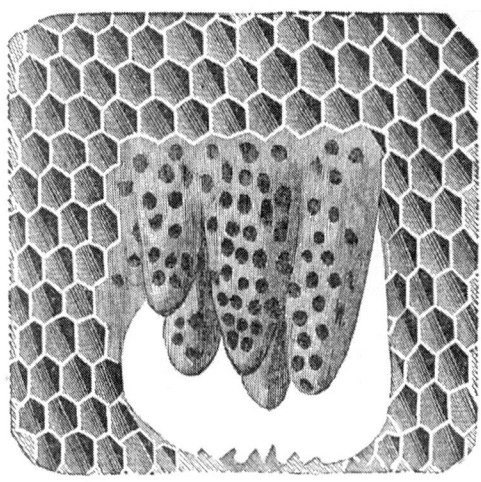

FIG. 7. The old way of having cells built.

beeswax and rosin) and so arranged that you can place a lamp under it to heat it. Keep in this pan a mixture of two parts rosin and one part beeswax. Heat this enough to work well, being very careful not to overheat it, as it will destroy the eggs in the cells if used too hot, and if too cold it will not adhere properly ; the right degree of heat will soon be learned

[5] Represents a cluster of cells built by the old method; a piece of comb containing eggs was inserted and as none of the eggs were destroyed the bees built cells in clusters as shown by this cut. The cluster shows five cells three of which may be saved by transferring.

by practice. I keep on hand a number of frames of comb which are free from brood or eggs, into which the prepared strips A A are fastened, as shown in fig. 6, page 15. You will notice that this comb is cut with a slightly convex curve. By putting the prepared strips in after this manner, still more room is given to each cell owing to the spreading caused thereby.

HOW TO FASTEN THE PREPARED STRIPS IN POSITION.

To fasten these strips, dip the edge which has not been cut into the preparation previously mentioned, and immediately place it in position, so that the mouths of the cells point downward, pressing it with the fingers gently into place, taking care not to crush or injure the cells in so doing. A number of such combs, say eight or ten, and more if necessary, should be kept on hand for this purpose and used as often as cells are needed. This is a great convenience and saves destroying or mutilating other combs. When the cells are cut out, the strip on which they are built should be taken with them.

HOW TO RELEASE THE BEES FROM THE SWARMING BOX.

Everything is now ready for the queenless bees in the box, impatient to be released and anxious to commence cell-building. This box has a strip of tin nailed on the upper edge of one end flush with the

2

outside; the cover has a similar strip nailed on the under side, which corresponds with the one on the box when in place, fig. 2. (See description at end of volume).

The combs containing the brood prepared for cell-building are now placed in the nucleus hive with other combs sufficient to fill it. Place this nucleus on the stand from which the bees were taken, and at such an elevation that the bottom edge of the alighting board will come just even with the top of the box in which the bees are confined. Next let the bees out by drawing the cover back just enough to allow the worker bees to pass between the strips of tin into the hive.

HOW TO SEPARATE THE DRONES FROM THE BEES.

If there are drones with the bees either black or otherwise objectionable, they will be retained in the box and can be easily destroyed, after the bees have all left. In case the drones are needed (or if there are none with the bees which is often the case), all may be turned out in front of the nucleus, when they will soon run in; this being on the old stand they will accept it as their home and begin cell-building at once from the eggs given them. In no case should any worker brood be given to the bees, thus compelling them to concentrate their whole forces on cell-building. Some capped drone-brood should be given them, if at hand, as it would greatly encourage the bees, and is really an advantage.

WHEN TO PREPARE THE BEES FOR CELL-BUILDING.

I usually prepare the bees in the morning for queen-rearing, and give them the eggs at night. By the next morning they will usually become reconciled to the new state of things and from twenty to twenty-five queen cells will be started; this of course depends upon the number of eggs given them. Just here let me caution all queen-breeders against giving the bees too many eggs, or allowing them to build too many cells at one time. If not permitted to complete over twelve cells, the queens will be found as good as, if not superior to, those reared under the swarming impulse. If you wish to rear queens of which you and your customers will be proud, you will find the secret is in not allowing any colony to build more than this number. If this precaution is taken, good queens will be the rule and not the exception. Of course, to accomplish this your breeding queens must in all cases be as near perfect as possible, other things being equal. We seldom find more than twelve first-class cells, and often a less number, in a colony which has just cast a swarm, the Cyprians and Holylands being exceptions. The queens of these races are very prolific and hardy, consequently they rear a much larger number.[6]

Now we have everything in good order and condition, and cell-building is progressing satisfactorily, except perhaps too many cells are being built for the

---

[6] This point is further described under another heading.

quantity of bees in the colony. If such be the case, and it does sometimes so happen, all over twelve or fifteen should be destroyed. This may be done by means of a match, as before stated. Sometimes, in the hurry of preparation, an egg is passed by and not destroyed, but if the work in the first place is properly and thoroughly done no trouble of this kind need be apprehended.

CHAPTER IV.

FEED WHEN FORAGE IS SCARCE.

If the honey harvest is abundant during the period of cell-building little care is needed until the cells are about ready to be cut out, but if not the nuclei

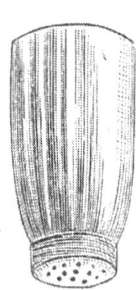

FIG. 8. Fruit jar feeder.

will need liberal feeding (say one pint of syrup each night and morning) to stimulate them properly. In such cases, feeding should certainly be resorted to or inferior queens will be the result. For this purpose I never have seen a feeder so convenient and so good as one constructed as follows: take either a quart or pint improved Mason's fruit jar, fig. 8, remove the glass top and substitute a tin one in its place, cutting the tin the exact size of the glass top and punch about twenty small holes therein for the food to pass through. Bore an inch and a half hole in the cover of the nucleus, and place the jar bottom upward over it. There should be a space of one-half an inch between the jar and the top of the frames so that the bees can get at the food readily. The bees will take a pint of food in the course of two or three hours, if there are twenty or more holes for it to pass through. For slow feeding make about six holes. If honey is scarce, use granulated sugar and water, say five pints of water to six pounds of sugar, and mix either hot or cold; make a fairly thick syrup. When thoroughly dissolved, flavor with a little pure honey. *Do not use glucose or grape sugar under any consideration.*

WHAT TO DO WITH SURPLUS QUEEN-CELLS.

At this point we will consider that the bees have been at work four days on the cells, and that they are sealed over or nearly so. If desirable they may be left seven days longer where they are, and then cut out and either placed in nuclei or put in the queen nursery to hatch.

Where queens are reared on a large scale, the combs containing the cells just sealed may be taken from a number of nucleus colonies and given to one, as one colony can take care of one hundred cells as well as a smaller number. Care should be taken to have a colony on hand prepared to care for such surplus cells until it is safe to remove or cut them out.

It will be necessary to give such colony a frame of

brood occasionally, to keep it well stocked with bees. These combs should be examined at least once a week in order that no queen-cells are built, as a queen might hatch out some day very unexpectedly and destroy all the cells in the hive. You will remember that each frame containing cells has the number of the breeding queen and the date of starting the cells marked on the top. If proper care is taken to keep a correct record of this in a day book kept for the purpose, you will know exactly when the cells are ready to hatch and the time they should be transferred to nuclei or to the nursery.

CUTTING OUT QUEEN-CELLS.

When the cells are sufficiently matured to be safely removed from the nuclei, cut them out, taking with them the strip of comb on which they were built. They should be immediately taken into a warm room and separated. A lighted lamp is kept at hand with which to warm the knife. Occasionally a small piece of one of the cells is shaved off. The fracture thus made may be easily repaired by placing a small piece of foundation over it, plastering it on, having the knife quite warm, being very careful to make sound work of it or the bees will reopen the cavity and remove the nearly matured queen. After this is done place the cells in the nursery or nuclei.

We will suppose that we have one hundred queen-cells on hand and one hundred nuclei ready to receive them. When the cells have been sealed seven days they may be safely cut out, but it is better to let them remain until the morning of the eighth day after they were sealed, and then give them to the nuclei. Of course, cells are being built in other hives and will soon hatch. The first lot of queens should be fertilized by this time, and disposed of, if the weather has been favorable, and room made for a second lot; but supposing the first lot not to have been fertilized on account of unfavorable weather, and there is not room for the second lot in the nuclei already in oper-

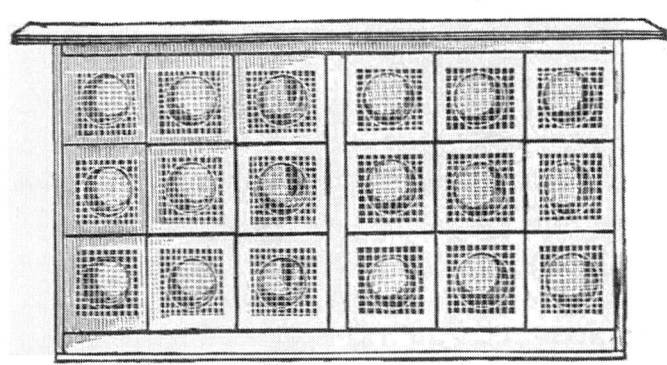

FIG. 9. Queen nursery.

ation, what shall we do with them? No queen-dealer can afford to lose a fine lot of cells, especially if he has a large number of orders on hand; every cell and perfect queen must be preserved in some way. How can this be accomplished? I will give my method which is a simple and good one. I provide against such a contingency by having a queen nursery of my own invention, fig. 9.

QUEEN NURSERY AND HOW TO USE IT.

I use eighteen cages in one standard Langstroth
frame ; each cage, fig. 10, has a place in it for a sponge
to contain the food, and another for the cell. An inch
and one-half hole is made in each cage, both sides of
which are covered with wire-cloth. Each cage is provided with food sufficient to last a queen one week.
The cells are cut out and placed in these cages which
are then placed in the centre of a full colony. As
eighteen of these just fill a frame they will stay in

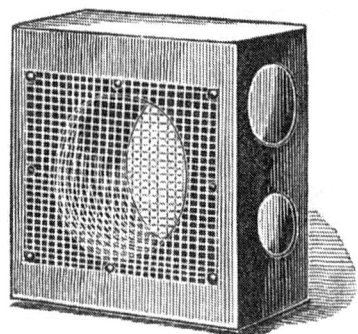

FIG. 10. Nursery cage.

place without any fastening whatever. A bungling
workman cannot make them so that they will work
just right. The cages must be exactly alike, and
then there will be no trouble in having them stay
in place. They should be cut out with a circular
saw run by steam power. I make it a point to do
all such work in the best manner possible.

By the use of this nursery, my queens are all hatched
in the brood-chamber and in nearly the natural way,
and by the natural warmth of the bees; no lamp

nursery nor other artificial devices being used, and none of which ever should be used in queen-rearing. When one digresses from the paths of nature, in this business, the more unsuccessful he will be. When the queens are in the nursery in the centre of the brood-chamber, they are perfectly happy and contented, and will live there safely for several weeks. In no other way have I been able to preserve them so long. Each cage must be supplied with food, as a colony with a laying queen will not feed virgin queens, and oftentimes even a queenless colony cannot be depended upon for doing so, as I have learned to my sorrow.

INTRODUCING VIRGIN QUEENS ; HOW LONG TO KEEP A COLONY QUEENLESS BEFORE INTRODUCING A QUEEN.

This is another important point which properly comes under the head of queen-rearing. It has been said that virgin queens cannot be successfully introduced. Those who assume this certainly mistake. I am obliged to introduce hundreds of them every year, and have no trouble in so doing. I seldom have occasion to introduce them to full colonies, but that it can be safely performed I have no doubt.

In order to introduce such queens successfully the colony should remain queenless three days (seventy-two hours) ; then give them a pretty good fumigating with tobacco smoke. Remember, the bees must remain queenless *three* days at the least, and during the meantime no queen must be near them, otherwise the operation will prove a failure. Virgin queens can also be introduced successfully by daubing them with honey and using no tobacco smoke. Put a little honey into a tea-cup and roll the queen in it so as to daub her thoroughly, then drop her from a spoon into the hive among the bees. They will at once commence to remove the honey and when they have done so the queen is safely introduced. This is a much slower process than by fumigating them with tobacco smoke, but quite as successful. Do this just before sunset. When tobacco smoke is used to introduce them, throw some grass against the entrance to keep the smoke in and the bees from coming out. Blow in a liberal amount and then let the queen run in at the top through the hole used for the feeder.

ANOTHER WAY TO INTRODUCE VIRGIN QUEENS.

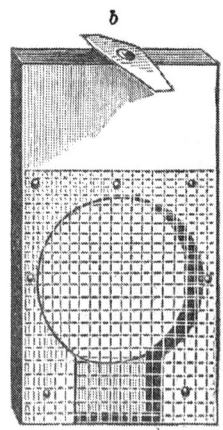

FIG. 11. Introducing cage.

Make a cage wholly of wire-cloth or such an one as is used in mailing queens. Cut a mortise from the main apartment to the outer edge as seen in fig. 11, cage the queen and fill the mortise with Good's food; by the time that the bees have removed it, they will have become acquainted with the queen. Bear in mind that the bees must be queenless three days before introducing virgin queens. If a little tobacco smoke is used to scent the bees at the time the cage is put in, I think the undertaking will be rather more successful. Laying queens can be introduced by the same process. A colony made queenless for the purpose will always begin to rear a number of cells. When the new queen is introduced, they immediately stop cell-building; all are destroyed, and the bees commence to pay royalty to the new queen given them.

CHAPTER V.

THE OLD AND NEW METHODS OF CELL-BUILDING.

I presume the reader has followed the descriptions carefully and understands now how to have queen-cells uniformly built, so that none need be destroyed in cutting them out. By the old method a large number of fine cells must necessarily be destroyed in transferring them, as they are built so irregular and oftentimes so close together that three or four are rendered useless. It is also exceedingly difficult to determine when the cells will hatch, as the bees will use eggs or larvæ in various stages from which to rear queens. Again the bees will continue to start cells three or four days after the brood is given them. My method does away with all this trouble, and we can determine to a certainty, within a few hours at least, the time when the cells will hatch. This is one of the great advantages to be derived from its use, and again, there is no guesswork about it, and no eight or ten day queens need be anticipated as none can be reared, simply because there are no old larvæ for them from which to rear queens.

HOW TO REAR QUEENS IN FULL COLONIES AND STILL HAVE NONE QUEENLESS.

I have asserted that my queens are reared in full colonies and none are ever queenless. Well, none of my standard colonies are ever without queens.

I first determine as nearly as possible the number of queens I intend to rear during the season, and then select enough good strong stocks for the purpose. One half-dozen colonies will rear a large number of queens in a season, as will be seen by what follows. I deprive the first half-dozen colonies used for queen-rearing of all their brood and queens, giving the former to weak colonies to build them up, as by this process I can soon make them strong. I prepare as many colonies in this way as I wish to keep building cells at one time and as I think will meet my wants for one season. I would state here that no colony should be permitted to build two lots of cells; I mean by this that the first lot of cells should not be removed and eggs given them to build others. I consider it poor policy to do so, although such a colony will build a second lot of cells; but they would produce inferior and almost worthless queens, and a queen-dealer who would do so would soon lose his reputation.

HOW TO OBTAIN MORE CELLS.

When I want a fresh lot of cells, I take a strong colony and remove all the bees, by the process described on page 8. Replace the combs, and put the hive exactly where one of the colonies has been standing which has just finished a lot of cells. Then brush or shake the bees from the combs of the latter hive (care being taken not to shake the comb on which the queen-cells are built), in front of the hive from which the strong colony has just been taken, and give them the queen from the first colony.

Bees enough should be left with the cells to keep them from chilling, or the frame on which they are built can be placed in a hive where other cells are being built, or add one frame of honey and one of brood, and form a three-frame nucleus. When the cells are ready to hatch, transfer all but one to other nuclei, and thus gradually form the needed nuclei for the season.

HOW OFTEN THE SAME COLONY MAY BE USED FOR CELL-BUILDING.

Having taken care of the cells and brood in the full colony, we now have a fresh lot of bees on our hands, which will be ready in a few hours to build other cells, while the bees which have just completed the first lot will continue the work in the hive from which the fresh bees were taken. In three or four weeks the same process can be repeated, as the old hive will then be full of newly hatched young bees. After getting bees enough to start the first half dozen lots of cells, no more hives need be made queenless and every colony will be kept strong during the season, as they will have a laying queen all the time. It will be seen by any intelligent bee-keeper that it will not require half as many colonies to rear one thousand queens, as by the old process. This I also claim as original.

DESTROYING DRONE BROOD AND WORTHLESS DRONES; THE USE OF THE DRONE TRAP.

Where the extractor is used drone-brood not needed may be destroyed after the honey has been slung out, by uncapping it, being careful to shave off the heads of the drones; for this purpose I use a thin, flat knife such as Mr. Peabody sold with his extractor. Excessive drone-breeding can be kept down by such a process, when there are only a few hives kept. I cannot recommend this process in a large apiary and some other device must be resorted to.

Most any one, I think, has sufficient ingenuity to construct a drone trap for the purpose of destroying worthless or surplus drones. I find it rather difficult to describe one so that all may understand it, but shall have an engraving made and give a description of it at some future time. All that is needed is to place a gauge at the entrance so that the drones cannot get out, and make an outlet for them to pass through into a box from which the workers can escape and they cannot. The swarming box, only on a smaller scale (of which a description is given at end of volume), will answer every purpose. Care must be taken, if a trap is used, not to smother the colony; this will not be the case if the bees have an easy means of escape from the hive into the trap. The outlet for the drones to pass into it must be made large enough for them to pass through freely.

HOW MANY QUEENS OUGHT A COLONY TO REAR?

I have frequently cautioned queen-dealers and those who rear queens simply for their own use, against rearing too many queens in one hive at the same time. This is so very important that I must be excused for repeating it. By my plan one hundred queens can be raised by a colony as well as twenty-five; but the more queens reared the poorer they will be. The correct number, as my experience teaches me, is about twelve queens to a colony. I have found that worthless queens are reared under the swarming impulse as well as by artificial means. If a queen is removed from a full colony they will build from twelve to twenty queen-cells. Very few of these will prove to be valuable queens, as a colony thus made queenless will not start cells from the eggs. They will select larvæ two or three days old for cells as their impatience leads them to diverge from nature's course every time. The queens reared from the latter cells always prove short-lived and almost worthless. I am aware that a large number of the queens reared are produced by the

latter process. This statement is founded on the
writings of many who rear queens.

HOW TO REAR VALUABLE QUEENS.

To rear valuable queens, I think the hive should
not contain over twenty-five eggs to start with, and
in about twenty-four hours after the cells are started,
at least ten of them should be destroyed, so that not
over fifteen remain to mature. Of course we cannot
afford to sell such queens for seventy-five cents or
one dollar each. Out of twelve cells that hatch, probably four of that number would never mature or they
would be destroyed in some other way : lost in mating,
killed in the hive, or by some other mishap to which
queens are always liable and exposed.

CHAPTER VI.

HIVE TO USE FOR YOUNG QUEENS.

For many years I have used the small hive described
on page 4, for queen nuclei (for fertilization only),
and find them as convenient and handy as any. They
are made large enough to take four combs, but I seldom use over three to each hive.

Many large queen-dealers use the same kind for
this purpose, and find them just as good as larger
ones, much handier and less expensive. In the fall

the bees are united with other colonies, and the combs packed away in barrels for use another year.

HOW TO FORM NUCLEI.

To prepare these hives for the cells, or young queens, I proceed as follows :—if a box-hive is to be broken up, with which to fill them, it is taken into the bee-room. The bees are treated the same as though they were in a frame-hive, viz. : induce the bees to fill themselves with honey by closing the entrance, smoking them, and rapping smartly on the hive for ten minutes or more. After driving all the bees out that can be induced to leave the hive, proceed the same as in transferring, placing the bees in the cap of a hive or box until ready to be placed in the nuclei.

Put the comb containing brood and that containing honey into the small frames. Give each hive at least one frame of brood, one of honey and one empty comb. Place two of the frames in the hive then put in about one pint of bees, putting the third comb in last, then place the cover on. Confine the bees in the hive forty-eight hours before permitting them to fly. If there are but few drones in the hives and we wish to destroy them, it can be quickly and easily done by the following plan: go to each of them early in the morning of the day on which they are to be liberated, take an empty hive with you, give the bees a small dose of tobacco smoke, let them remain quiet a few moments then examine the combs separately and very carefully, and pinch the head of every drone found. Put the combs with the adhering bees into the empty hive, placing it on the stand of the hive just examined. The bees must have plenty of air while confined in these small hives. For this purpose I use a small screen to nail on the front of each, fig. 12.

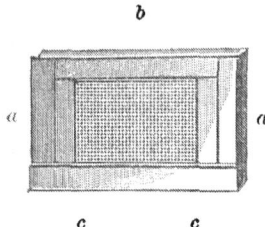

FIG. 12. Screen used for ventilation.

It should be made roomy enough so that the bees can come out from the body of the hive into it, and get all the air they need and return as often as they choose.

WHERE TO PLACE NUCLEUS HIVES.

Nuclei used for fertilizing queens should not be placed very near colonies building cells, as the queens, when returning from the marriage tour, are quite liable under such circumstances to enter the wrong hive, and young queens, even though not fertilized, are always welcome to a queenless colony. Nuclei for fertilizing queens should not remain queenless long at a time; if they do there is great danger of fertile workers gaining possession of the colony and they are the pests of the apiary. I have seen many valuable queens destroyed by them.

WHY BEES BALL AND DESTROY QUEENS.

When queens have returned from the flight in

search of drones they are sometimes seized by the
bees, hugged or smothered to death (called balling),
stung or injured, wings torn, or one leg stung and
rendered useless. In almost every instance where
this hugging takes place the queens are nearly ruined,
this being more than they can endure without injury
to their fertility. When this takes place one may
know that fertile workers infest the colony.

HOW TO INSERT QUEEN-CELLS.

Having everything properly prepared we are ready
to give to each nucleus a queen-cell. This can be done
without taking out the combs or cutting them, as is
the general method practised by most queen-dealers,
and given by the bee journals as the best. I
generally find plenty of room between the combs
without disturbing any of them. If not, I slip one of
the frames a little to one side, place the cell in position point downward, of course, and gently press the
comb back against it. It will thus be held securely
in place and will hatch as well as though inserted
in the centre of the comb. By doing this, the
combs are not mutilated and the operation is quickly
performed.

If young queens are to be given to the nuclei
instead of cells, proceed according to directions given
on page 25. Bear in mind that very young queens
are more acceptable than those four or five days old.
It is much more troublesome and more difficult to
introduce older ones, and the latter will be destroyed

unless scented by being fumigated with tobacco smoke or by some other means, the idea being to deceive the bees, which can be done by scenting them all alike.

A queen-cell can be given to a colony immediately after removing a queen from it, and should the young queen emerge from the cell within an hour she will generally be kindly received, and thus safely introduced.

Occasionally, a queen hatching so soon after the cell has been introduced will be killed, but this seldom happens.

Cells may be given to queenless colonies at any time, but queens should not be given to any colony until it has been queenless three or more days.

THE AGE AT WHICH QUEENS ARE FERTILIZED.

We read quite frequently in the bee journals of queens becoming fertilized when only three days old. This may be true, but in all my experience, I never knew one to take her wedding flight when less than five days old. In from thirty-six to forty-eight hours after this they usually commence to lay.

Early in the season they generally come out, between the hours of one and three P. M., and sometimes as late as four P. M. After the first of September they will fly as early as 11.30 A. M., and not much later than two P. M. unless the weather is very warm and pleasant.

HOW TO FORCE THE YOUNG QUEENS TO FLY.

In localities where forage is scarce, some means must be adopted to stimulate the bees and cause the queens to fly when they are not disposed to do so. This can be accomplished by feeding the bees. The nucleus feeder fig. 13, which I have used for twenty years, will hold one ounce of syrup and is admirably adapted for this purpose. Such colonies as have queens old enough to fly are fed during the forenoon and the queens will fly

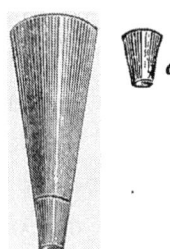

FIG. 13. Cone feeder.

in the afternoon and generally be fertilized; whereas if they are not fed they will not leave the hives sometimes until a week later.

QUEENS FERTILIZED BUT ONCE; HOW FAR TO KEEP THE RACES APART TO INSURE PURITY.

I am satisfied that no queens are fertilized more than once. They may fly more than once, but if they make the second flight and bear all the indications of having met a drone, it is pretty certain that they were unsuccessful the first time.

There are some who entertain the idea that a race of bees cannot be kept pure unless they are kept isolated several miles from all other races. I have tested this matter carefully and pretty thoroughly during the last twenty years, and have found that one-half mile is as good as a much greater distance.

While drones will sometimes fly a mile or more, the queens will not. This fact can be ascertained by watching a queen when she takes her wedding flight. She seldom is gone over five minutes and nine out of every ten will return within five minutes. Of course if the young queens are only one-half mile from a large apiary of black bees, there must be an abundance of Italian drones in the same yard with them. This being the case not one queen in twenty-five will mismate. This rule applies equally well to all races.

HOW TO KEEP LAYING QUEENS.

Sometimes queen-dealers and often other apiarists have occasion to keep laying queens on hand several days before using them.

Surplus queens can be kept on hand a long time, in the queen nursery, provided the cages are well supplied with food; this being the only attention needed. The sponges in the cages must be refilled with honey as often as once each week at the least.

The nursery should be placed in the centre of the brood-chamber of a full colony. To make room for it either remove one of the outside combs, or, in case the colony is strong, one of hatching brood. The latter may be given to some weak stock. When the nursery is taken from the hive and no other queens are at hand, fill the space then left vacant either with an empty comb or a frame filled with foundation.

A few queens may be kept by placing the cages between the cushion and frames, in such a manner that the bees will have easy access to them. A colony having a fertile queen cannot be depended upon to feed other queens under such circumstances, hence the importance of supplying the cages with food. A person must use his judgment regarding keeping queens in this way, when the nights are cool. There need be no apprehension regarding this point between June 1 and Sept. 1.

FEEDING NUCLEI; WHY THEY SOMETIMES SWARM OUT.

Nucleus colonies (in hives, described at end of volume), must be fed as often as once each week, unless forage is abundant, or they will "swarm out," even when they are well supplied with brood and honey, as such colonies are easily discouraged.

They will not do so if fed a small amount of syrup occasionally. Use only the best sugar-syrup (not honey), giving it to the bees the same day on which it is prepared. Never use poor food as it soon sours, and runs out of the feeder besmearing the bees and combs. This will surely cause them to desert the hives.

The following incident well illustrates this point: one morning I fed fifty nuclei with some syrup which had been prepared but twenty-four hours. The weather was very warm; the syrup became sour and ran out of the feeder besmearing the bees and combs, every colony of which swarmed out and united in one cluster. This resulted in a loss of nearly fifty dollars to me.

CHAPTER VII.

UNPROFITABLE BREEDING QUEENS.

I have found that many of the young queens from some mothers are lost on their "marital tour," although such queens when successfully mated prove very valuable. This is a singular and unnatural phenomenon for which I cannot account. Why they fail to return to the hive is more than I can comprehend. Queens having this imperfection should be discarded as breeding queens at once, notwithstanding they may in many other respects be very desirable as queen mothers. We cannot afford to use queens of this class from which to breed others, when so many of their young queens are lost in mating. It is rather discouraging when examining a lot of nucleus colonies, where there should be a number of fine laying queens, to find none, they having been lost while on their first flight.

MOST PROFITABLE BREEDING QUEENS.

I bred from a queen last season not one in fifty of whose royal daughters was lost in mating. These are the only profitable ones from which to breed. For breeding queens select carefully only those which are very prolific; whose royal progeny are fair in size and handsome, whose worker bees are uniformly marked, gentle, good workers, and perfect in all other respects. Purity of stock cannot be maintained unless great care is taken in selecting the queen mother. Never use one whose workers have from one to three bands. The young queens from such an one would show a variety of markings, black, striped, and a beautiful yellow, the yellow ones being sadly in the minority. If beauty, purity and business qualities are desired, such a queen would be worthless. Always select one having the markings which show her to be pure, prolific and hardy. This subject is more fully explained elsewhere.

THE SELECTION OF THE DRONE MOTHER.

I have long contended that success in queen-rearing depends largely upon the drones used for mating. The same care should be taken in the selection of the drone mother as with the queen mother. Her worker bees should be well-marked, gentle, good honey-gatherers, hardy, and absolutely pure; the drones large, handsome, and very active. I never permit drones from all my colonies to fly promiscuously, or have any haphazard mating of queens. Only selected drones having the above-mentioned qualities are tolerated in my fertilizing apiaries.

HOW TO REAR AND PRESERVE DRONES.

It is well known to most bee-keepers that colonies having fertile queens will neither rear nor permit drones to live in the hive late in the season, and sel-

dom when forage is scarce. If queen-rearing is going
on, drones must be procured at any cost, and some
means must be adopted to rear and preserve them for
use in the latter part of the season. To do this I
pursue the following method : have at hand several
extra frames of drone comb ; insert one in the centre
of the colony from whose queen you wish to rear
them. Feed this colony liberally if forage is scarce.
Examine them in the course of a week ; if the comb
is well filled with eggs and larvæ remove it to a
queenless colony.

Instinct teaches queenless bees the necessity of
rearing and caring for drones, hence they can always
be depended upon for them provided the brood is
given them.

Replace the comb just removed with an empty
one ; continue this as long as the queen can be induced to lay drone eggs. Remember that queenless
bees never destroy drones, while a colony having a
fertile queen will invariably do so, unless encouraged
to preserve them by being fed. It is a very difficult
matter here in the north to induce queens to lay drone
eggs in September, even when forage is abundant ;
hence drones to be used in September and October
should be secured in the early part of August, as
most colonies seem disposed to rear them at this time.

HOW TO JUDGE OF A QUEEN BEFORE TESTING.

The practised eye of an expert in any vocation can
detect imperfection where the novice cannot. My

experience enables me to judge of the qualities of a queen; whether she will prove prolific or otherwise, as soon as she has laid several hundred eggs. The laying queen, if a good one, will deposit all her eggs in the cells in exactly the same position. Every egg will point downward, and will be large and plump when com-pared with those of an un-prolific one, and every cell

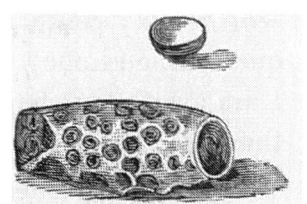

FIG. 14. Cell from which a strong queen hatched.

not otherwise occupied will contain an egg. I can also determine whether a queen is worth preserving or not the moment she leaves the cell. An inferior queen in gnawing through leaves a little ragged hole instead of cutting a large clean cap, fig. 14, and leaving an opening nearly large enough to admit the end of the little finger, as a strong and well developed queen always does. It is worse than useless to pre-serve such queens.

DESTROY WEAK AND FEEBLE QUEENS.

After the cells in the nucleus or those in the nur-sery are hatched, examine the outlet to each, and if small and ragged destroy the queens at once. A good prolific queen will lay nearly or quite four thou-sand eggs in twenty-four hours. One that will not do this is not worth preserving. I never saw a queen that I considered too prolific for my own use (the opinions of some others to the contrary notwithstanding). I want queens that will deposit ten thousand eggs or even more in a day if they can be reared.

Very few bee-keepers are troubled with queens being too prolific. I should as soon find fault with my hens for laying two eggs per day when eggs are worth forty cents per dozen, as with a too prolific queen when honey is worth twenty cents per pound. I am aware that hens do not eat the eggs after laying them as the bees do the honey after gathering it; but if the hens laid no eggs there would be no profit, and if a queen is not prolific there is no income from that source.

The main object in rearing queens is to get hardy and prolific ones, the more prolific the better. A queen that will lay five thousand eggs in a day is worth one hundred that will lay but one thousand in the same time. My opinion is founded on experience and the result of careful experimenting, and I believe that a large majority of bee-keepers and those of extended experience are of the same opinion.

LARGE *versus* SMALL QUEENS.

I find customers occasionally who want large queens. A person engaged extensively in queen-rearing will have them of most every size.

I must confess that I like the appearance of large and handsome queens; but they do not as a rule prove to be the most prolific or profitable. Queens of medium size are generally the best. They have proven so with me. Good queens are those that keep their hives well filled with bees. The color or size has no effect on their fertility.

An experienced bee-master can judge of the quality of the queens which he is rearing even before they leave the cell. If the cells are short and blunt when just sealed they should be destroyed at once, rather than wait and destroy the queens after they have hatched, as such queens would prove worthless.

The cells containing good queens are long and pointed, and heavily waxed with a rough surface. The bees when constructing the cells seem to understand the condition of the embryo queen, and whether she will be strong and vigorous or otherwise.

Cells having the blunt point and thin walls, so thin in fact that the young queen can be seen through them, generally contain very poor ones. Queens that delay long before becoming fertile in favorable weather will not prove of first quality. A smart, active queen will invariably leave the hive on her wedding trip when she is five days old, and in all my experience I never knew one to become fertile at a younger age.

BEST BEES FOR QUEEN-REARING.

A person rearing queens extensively, and keeping several races of bees in his apiaries, should note those that build the largest and finest cells, and rear the best queens.

The Holyland bees will build a large number of cells if permitted to do so, and the queens reared by

them are very large and prolific. All things considered, I believe them to be the best of the four races
for cell-building. The Cyprians come next in value
for this purpose; there is, however, little difference
between these. My acquaintance with the Cyprian
and Holylands, however, has not been as extensive
as with the Italian and black bees, and I will give my
experience regarding Italians, as it may differ from
that of others.

THE ITALIANS NOT A DISTINCT RACE.

The fact that the Italians are not a distinct race is
well established and generally admitted; hence it becomes necessary, in view of this, to propagate the
other races in order to keep the former up to the
standard and maintain their established reputation as
a superior race.

I have tested the different races and find that the
Italians are the least adapted for cell-building. I do
not understand why this is so. Prior to the introduction of the Cyprians and Holylands, I always
gave the black bees the preference as nurses, when-
ever I could procure them.

The cells built by the Italians are small, many being
similar to those described on page 45. I would not
hesitate for a moment to destroy them as soon as
sealed. It would be folly to permit queens from
such cells to become fertile, as they would prove
worthless, and much valuable time would be lost in
testing them. Distinctly remember that rearing

queens artificially is quite a different process from that pursued by the bees when allowed to follow their natural instincts. The Creator has instituted perfect laws governing insects, and bees comply with the requirements of these laws when allowed to rear queens in the natural way; the result being perfect queens.

In view of this and on account of the increasing demand for queen-bees, it becomes necessary to adopt some artificial means by which equally as good queens can be produced. In all my experimenting with bees for this purpose, I have imitated nature as perfectly as possible.

REARING QUEENS FROM THE EGG.

Most intelligent and experienced bee-masters agree with me in the opinion that queens should be reared from the egg, as, other things being equal, they prove the best in all cases.

I formed this opinion during my first year in the business, have had no reason to change it, and claim that this is the only way to rear queens which will compare favorably with those produced in the natural way under the swarming impulse.

Many years ago, I frequently heard an old bee-keeping friend make the following remark : "If you want large queens start them from the egg." Anyone can satisfy himself of this fact by thoroughly testing it for a few weeks. Full directions for procuring and keeping a continuous supply of eggs for this purpose may be found on page 4, which should be carefully followed in order that you may be prepared to start cells at any time.

this purpose may be found on page 4,

THE PROPER TIME TO COMMENCE QUEEN-REARING.

I make it a point to have the hives well stocked with bees and in a proper condition to swarm, having sealed drone-brood before I start my first lot of cells. Here in New England, in favorable seasons, queen-rearing may be commenced the first week in May provided the bees are properly stimulated.

It is well understood that natural cell-building depends upon the following conditions, viz. : strong and populous colonies, a good supply of drone-brood and young drones, vigorous queens, warm and genial weather, and a plentiful flow of honey. Hence, you will see the necessity of stimulating the bees for early breeding by giving them a liberal supply of syrup, and in every other possible way.

I consider the above the best indications of their readiness to commence cell-building.

CHAPTER VIII.

THE NEW WAY *versus* THE OLD.

I wish to point out some of the advantages which the new method has over the old. The latter having

been fully described in the various bee journals and
standard works on apiculture, the reader must necessarily be familiar with them.

The first point is to make a colony queenless. A
few hours later, give it a comb containing eggs and
larvæ (either with or without holes cut in it), permitting the bees to build the cells where they choose.
They generally build them in clusters (see fig. 7),
and so closely joined that they cannot be separated
without destroying many of them. The eggs given
vary in age from one to three days, consequently
when the queens commence to hatch, it will be from
one to three days before they have all left the cells;
and when a larva three days old is selected, the queen
will hatch in nine or ten days. Unless carefully
watched this early hatched queen will destroy the
remaining cells by gnawing through them near the
base, and, when she has made an opening of sufficient
size, will deliberately sting the imprisoned queen, the
bees finishing the work thus ruthlessly begun. I
used quite frequently to find five or more queens in
the hatching-box or in the nuclei at one time, and of
course many of them would be stung.

HOW YOUNG QUEENS MANAGE, WHEN TWO OR MORE HATCH AT ONE TIME.

Perhaps my readers are not familiar with queen-
rearing, and do not know how the young queens
manage affairs when several emerge at the same time.
Where several cells are clustered together, as shown

4

in fig. 7, several of these queens are likely to hatch at the same moment. When this occurs and they meet, a mortal combat ensues ; the conqueror coming out of the struggle unharmed, and the other receiving a fatal wound is left unmolested to die, unless some merciless worker seizes her by the wing and attempts to drag her out of the hive. Please remember that worker-bees never destroy a cell that contains a healthy queen. In all my experience I never knew such a thing to occur. Every queen will be permitted to hatch unless attacked by a hostile queen. I am, however, fully aware that many writers state that the workers *do* destroy them. When breeding by the method just described, I was obliged to spend many sleepless nights watching such cells as they could not be separated, and they would almost invariably hatch at night. The lamp-nursery system has many advocates, but I have never tried this plan as I consider it contrary to natural laws. Cells need the natural warmth of the bees, and it is almost impossible, in the lamp-nursery, to keep the temperature even. Such means will produce poor, weak queens.

After long experimenting, I discovered *my* method as described in this book. With this came a wonderful change.

The eggs being of one age when the cells are started, I can accurately determine the time when they will hatch, and they all do so within a few hours of each other. The cells are uniformly built and can be easily separated. By the use of the queen nursery they are hatched by the natural warmth of the bees in the brood-chamber. The fact of knowing just when the queens will hatch, would have saved me hundreds of dollars had I known it when I first engaged in the business. Now I am not obliged to sit up nights to watch the cells and save the hatching queens, and I feel certain that my readers, without one exception, will admit that the above is entirely new, simple, and practicable, and see at once the advantage to be gained by its adoption. If the directions given are strictly followed, no queen need be lost when hatching, as the time can be calculated to within a few hours, as follows : the egg will hatch in three and one-half days after being laid ; four and one-half days later the cell will be sealed, and in eight days more the young queens will hatch out, making sixteen days, the time required to rear a queen from the egg just laid.

HOW TO REAR A FEW QUEENS.

As many bee-keepers desire to rear a few queens for their own use, thus combining pleasure and profit, I will give special directions for so doing. The general rules given in a preceding chapter should be followed, but all operations will necessarily be on a smaller scale. If you wish to rear about a dozen queens, go to a strong colony at sunset, remove the queen, and on the following night take away all the unsealed brood (replacing it with empty combs) which may be given to some weak colony ; then examine all the remaining combs, carefully destroying all cells which have been started. Now give them eggs prepared for cell-building in the manner described on page 12. Mark the date of starting on the hive. Five or six days later take a three-frame nucleus hive, and place two combs in it; now take the comb on which the cells are started, together with adhering bees from the first hive and place it in the centre of the nucleus hive between the two others; then select several combs of brood from strong stocks, giving them to the colony from which the cells were taken. Next, give them a queen, letting her run in at the entrance and she will be kindly received.

Or, in case we wish to rear other queens, remove the bees from another strong colony, replacing them with those that have just completed the lot of cells, giving them the queen just removed and proceed as with the former. It will be necessary to have one of the swarming boxes for the last lot of bees. The above directions apply only to full colonies and standard frames.

HANDLING QUEEN-CELLS.

This also is an important matter to those who rear queens. Great care should be taken in handling combs on which cells are built, as when the cells are not sealed the slightest jar may detach the pupa or nymph from its position and separate it from the jelly-food; and, although the bees may elongate the cells and save the queens, they will be permanently injured, but in most cases of this nature the cells will be destroyed soon after being sealed. The combs should not be handled except when absolutely necessary, and never tip or turn them bottom upward. In no case attempt to shake the bees from the combs, allow them to stand in the sun, or exposed in the cool air until they become chilled. Any rough or careless handling will result in injury to the embryo queens.

The wings of the young queens are not perfectly formed until within twenty-four hours of the time they hatch; and if the cells are subjected to such rough treatment, many of the queens will be crippled by having defective wings or legs, or perhaps the abdomen deformed. These precautions apply equally well when cutting out cells.

Queens may be hatched from those that have been chilled, but they will be weak and sickly; hence worthless.

In changing combs with cells on them from one hive to another, do not brush the bees from them, but let them remain to protect the cells from the extreme changes in the atmosphere. The bees adhering to such combs are kindly received by others under like conditions and circumstances.

Queen-cells should never be exposed to the burning rays of the sun as the cell in which the queen is encased is almost air-tight, and such exposure invariably produces suffocation and death. If the temperature in the room where the cells are being prepared for the nursery or nuclei is slightly lower than that of the hives, the cells will not be seriously affected by it, but do not keep them out of the hives longer than is absolutely necessary.

REASONS WHY QUEENS FAIL TO LAY, DIE SUDDENLY, OR ARE SUPERSEDED.

During the past ten years complaint has been made by some of my customers that queens sent them failed to lay after being introduced. Had these reports originated with unreliable parties, I should not have considered them worthy of notice, but on the contrary the complaints came from customers in whose honor and integrity I have the greatest confidence.

This led me to investigate the matter fully, as to whether queens taken from strong colonies while they were in a high state of prolificness and fertility were not more liable to injury in transit.

The matter was certainly of sufficient importance to demand thorough investigation. Occasionally, the purchaser would report: "My queen came to hand and was safely introduced, but has not laid an egg up to date." Of course I felt chagrined to hear a report like this from a customer to whom I had sent a *tested* queen.

When *dollar* queens are sent such a report will not surprise any dealer, as such queens are neither warranted nor tested; the only stipulation regarding them being that they are fertile, and little further is known of them as they are generally shipped as soon as they commence laying.

I was well satisfied that if these reports were correct, the injury must have been the result of rough treatment during transit.

In order to substantiate my opinion and conclusion, I was obliged to experiment considerably; consequently I removed, on different occasions, several queens from nuclei and full, vigorous colonies, keeping them in the nursery for a few days before shipping, also keeping a correct record of these queens and their destination in order to ascertain if they reported all right. No complaint came regarding them, hence I concluded that I had discovered one of the causes and also the proper remedy for it.

The above is not the only reason why queens fail to lay. Unless properly introduced, they will be rendered worthless before they have been in the hive an hour. Sometimes they will be slightly stung, but not sufficiently injured to cause immediate death, although rendered incapable of laying. When the hive is opened the queen is, apparently, kindly received by the bees and successfully introduced, as the marks made by the sting are not always easily recognized.

Occasionally they receive a sting in the leg rendering it useless, and such injury is easily recognized. Nevertheless, the queen will continue to lay, but not to the extent that she would had she received kind treatment from the bees when introduced. Sometimes, several weeks will elapse before they show any indications of failing or having been stung, and then are superseded, or as the term is "missing."

Parties purchasing queens should not hastily condemn the dealer, but should carefully study the causes of the loss. There are many reasons why queens die suddenly, fail to lay, or are superseded soon after being introduced, the principal of which have been described above with the remedy for the same, and regarding the others I can only say, use caution in introducing them. I would advise the reader (if a dealer) to keep queens "to be shipped," in the nursery for a few days before sending them out. I am satisfied that should this plan be generally adopted, less queens will be lost or injured in shipping.

CHAPTER IX.

TESTED QUEENS THE STANDARD OF EXCELLENCE.

This is an important subject and one of great interest, especially to every honey-producer. Queen-rearing has become a specialty, and honey-producers who wish to rear queens for their own use, and the queen-breeders who desire to infuse new blood to prevent in-and-in breeding want good reliable stock, and in order to obtain this call for *tested queens.*

I think that the importance of the term is often forgotten; some consider that it simply applies to a queen whose worker progeny bear the markings which indicate purity. If so, they mistake, and I think it due to the dignity of queen-rearing and api-culture that this matter be more clearly explained and definitely established.

There must be some *standard of excellence* and I consider that this is implied in the term *tested queens.* It is not my intention to detract from the value and importance of the business by devising means for producing cheap queens, but to establish a method by which the best may be produced, thereby promoting its interests and worth.

If there is one thing more than another which will degrade any business or profession it is trying to produce a cheap article. This invariably leads towards fraud and deception, and results in general injury and loss. Where is the dignity of the mechanic to-day as compared with the past? This principle of doing cheap work has ruined it, and it is almost impossible to get honest work done by contract.

Now with regard to queen-rearing, teach the mass of bee-keepers some way to rear queens cheaply in large quantities, and the country will be flooded with poor and worthless queens. First-class queens cannot be reared and sold for one dollar, and those who expect to get such queens for that price will be disappointed.

Tested queens are those bred from the best stock and kept in the apiary until the value of their progeny regarding honey-gathering and purity has been thoroughly determined; and no queen should be shipped as tested until the above mentioned requirements have been complied with. All tested queens should be graded, the best being *selected tested.* I have such an one in my apiary for which I should refuse one hundred dollars.

SELECTED QUEENS.

Selected queens are those which give promise of being perfect in all respects before being tested. When I open a nucleus hive and find a large, handsome and prolific queen, one that is putting the eggs in every cell in exactly the same position, I mark her either selected or for testing; and if she is shipped before she has been kept long enough to test her progeny she is called a selected queen. Such queens (all things considered) are the cheapest in the end for bee-keepers generally.

WARRANTED QUEENS.

My apiaries are so located and arranged that very few of the queens will mismate; hence most of the queens, sent out as warranted, prove to be purely mated; they are reared from the best stock and just as carefully as the tested ones. The chances are that the purchaser will be well satisfied with the warranted queens, especially if he be a honey-producer, although I consider the tested queens far preferable for those who wish to breed queens.

DOLLAR QUEENS.

As before stated, I regard the production and sale of *dollar* queens (so called) an injury to apiculture and its interests. I do not rear such, have none for sale, and advise the reader never to purchase them of any dealer.

ROBBING NUCLEI; HOW PREVENTED.

When nuclei are kept in the same yard with full colonies there will always be more or less robbing during scarcity of forage, especially when feeding is resorted to; and any apiarist who has once experienced wholesale robbing in his apiary will never forget it. To prevent this, feed only white sugar syrup as there is no "enticing scent" to such plain, simple food.

Clear honey cannot be safely fed, no matter how much precaution is taken, and should not be used even though the honey costs nothing and sugar fifteen cents per pound.

I know of nothing more discouraging to the queen-dealer (unless it be unfavorable weather) than upon examining his bees to find the strong colonies robbing their weaker neighbors; and as it is not easily controlled when once commenced, every means should be used to prevent it, as "an ounce of prevention is worth a pound of cure." There are, however, times when we shall find the bees robbing in spite of all our precaution; and there are several plans either of which may be resorted to in such cases, and unless this is done every nucleus in the apiary will be ruined in a short time. When a colony is being robbed, close the entrance at once to keep the robbers in the hive from coming out and others from entering. After giving the robbers confined sufficient time to fill their sacs, release them sprinkling them with flour as they leave the hive and trace them to their home. The hive being robbed should be closed and opened repeatedly until all the robbers have vacated. Then give the robbing colony a dose of tobacco smoke, which will soon stop their marauding (for a while at least), and when they have nearly all returned to the hive close it for a time with one of the screens, fig. 12, page 34 (described at end of volume), thereby checking and preventing further robbing.

If the hive being robbed is queenless and reduced in numbers, it should be removed to the stand occupied by a stronger one (also queenless), thus equalizing them. I find this method of equalizing nuclei a good one, even when there is *no* robbing.

The entrances of the hives should be protected with a piece of glass (four or more inches long and one inch wide) placed against them in such a way that the bees can pass out at either end, and secured against the hive with two small tacks.

If robbery is being carried on to any great extent, one end may be closed with a piece of paper; and if the robbers are still persistent, throw some grass against the glass in such a way as completely to cover the entrance.

The bees belonging to this colony will find their way in or out of the hive, while the robbers hesitate before forcing their way through this barricade; and if they should attempt it, their chances of escape are few, for it is not an easy matter for them to find their way out again, and the colony thus assailed takes courage, while the sentinel bees, with renewed vigor, seize upon the intruders as they enter, sting-

ing them before they can escape, often defeating
them. The glass placed against the entrance in no
way interferes with the queen or bees when they
wish to pass out or in. When feeding nuclei, great
care should be taken that the feeders do not leak,
and that they are properly filled. If any syrup or
honey is spilled upon the ground, cover it with earth
so deeply that the bees will not discover it.

FERTILE WORKERS.

One of the greatest and most troublesome pests of
the apiary (especially to queen-breeders) are the
fertile workers. They are generally produced by
allowing a colony to remain queenless for a long time,
appearing sooner in nuclei than in full colonies.
Their presence is known by drone-brood in worker
cells. Drones reared from these eggs are invariably
dwarfed, and, in my opinion, incapable of fertilizing
queens. It is quite difficult to introduce virgin
queens to such colonies, although a cell may be
safely given them at any time. The remedy is this:
place a frame of well matured worker-brood in the
centre of the brood-nest, and in a few days give
them a well matured queen-cell, and by the time that
the brood is all hatched the fertile-workers will be
gone.

When you think a colony is infested with them,
examine the combs, and if the eggs are laid in a careless manner, two, three, and often more in a cell,
with many cells passed by, and the brood when

capped projects beyond the worker-cells, you may be sure that it is some of their work. It would be well to destroy all such brood before giving the colony a queen-cell. Do not destroy the combs but rather shave the heads off with a sharp knife.

CHAPTER X.

ROOM FOR TRANSFERRING AND HANDLING BEES.

In most large apiaries we find bee-houses in which rooms are provided, adapted to this purpose. No apiary is complete without one. The windows in this room should be so arranged that all excepting one can be darkened. To show the advantage of such a room let us suppose that we are transferring ; all the bees taking wing during the operation will fly to the window which has not been darkened. After the work is finished, the hive should be returned to its stand and the window opened so that the remaining bees can return to it. If the bees were brought from a distance, the hives may be placed on stands prepared to receive them. In the latter case I leave the window closed allowing the bees remaining in the room to have their own way. They will cluster in one corner of the window and the next morning they may be brushed into a box and returned to the hive. Again, colonies containing objectionable drones may be taken into this room, the queens secured and caged, the bees brushed into the swarming box, the combs put back, the hive returned to its stand and the drones screened out by placing it at the entrance of the hive, retaining them and releasing the bees as described on page 18. Place the caged queen in the hive before releasing the bees and then release her as soon as they are quiet. Use as little tobacco smoke as possible during this operation, else the bees will be a long time in recovering from the effects and the brood will thus be unnecessarily exposed. In my business I am obliged to purchase black bees in large quantities — they are taken into the bee-room and the drones screened out, by the above process. I could not keep my bees pure unless some such means were adopted to get rid of them, and could not conduct my business properly without such a room. I have every thing conveniently and practically arranged in order to simplify all necessary operations; always keeping in view the fact that "system is one of the most prominent secrets of success." Again, one of the most important pieces of furniture in the room is the honey-bench, which should be permanently located in some convenient part of the room near a window, as we need considerable light when cutting brood into the frames. It should be six feet long and three feet wide, the front being three inches lower than the back, so that the honey may run down into a trough on the front, thence through a tin tube into some receptacle under the bench. The height may be arranged to suit the operator. It should be covered with zinc to keep the wood from absorbing the honey and in order that it may be cleaned easily; or, if zinc cannot be conveniently procured, take equal parts of beeswax and rosin, heat until hot, pour it upon the bench, spreading it equally and burning it into the wood with a hot sad-iron. The latter answers the purpose equally well. All pieces of comb containing honey may be broken up, thrown upon this bench and left until the honey is drained out; and all waste pieces of comb, left from this and other sources, should be thrown into the wax-extractor and the wax rendered from them.

The wire-cloth with which the windows are covered is tacked on frames which are hinged so that the windows can be opened to allow the bees to escape when necessary. These screens prevent robber bees from entering when forage is scarce and the windows are open to ventilate the room.

WHAT TOOLS ARE NEEDED.

It is necessary to have a few convenient tools at hand in order to do the work properly, including: a carpenter's shingling hatchet; a chisel with thin steel point tapering back two inches from the end for cutting off nails in the joints of box hives; a small, light, corn broom, such as is used for brushing clothes (I find this the best thing for brushing bees from the combs); a long, stout knife for cutting out combs and a small thin one (previously described) for cutting out queen-cells; a liberal supply of water and a towel which are very convenient for cleansing the hands, when transferring bees and extracting honey, especially if there is new honey in the combs; last, and not least, two smokers, the one a Bellows smoker, the other for burning tobacco. It is well to have a number of pieces of board (kept in some convenient place) to use during transferring, also some cotton twine for fastening combs into the frames. The above constitutes the necessary equipment in the shape of tools and fixtures with which to perform our manipulations in the bee-room properly.

TRANSFERRING BEES.

It is quite an undertaking to the novice, especially if he is afraid of bee stings, to transfer a colony to a frame-hive, and his troubles seem greater if the colony is a large one, and the weather quite warm, but when the operation is well understood, it is quite simple and easy to perform. The preparations for this work need not be very extensive. (The tools needed have been previously described in connection with the bee-room). If many colonies are to be transferred the following directions should be noted. Remove from the bee-room all unnecessary furniture. If the weather is cool the room should be warmed, so that the brood will not chill, and the bees can be handled much more easily with the temperature about 80°: it is a poor plan to handle them when below 70°. Take the colony to be transferred into the bee-room, smoke the bees and drum on the hive a little to cause them to fill with honey, and invert the hive on a table, box, or anything that may be convenient. While the bottom-board is being removed, use a little smoke to drive the bees down among the combs, and place a box over them, drumming smartly on the hive to drive all the bees out of it that can be made to leave. It will make very little difference to the expert whether the bees are first removed or not, but the novice will find it much more convenient to have them all out of the way, while removing the combs. Now take off the side of the hives from which the combs can be most easily removed, using a thin chisel to cut the nails in the joints before removing the bottom-board. The jarring thus caused answers the same purpose as drumming, and by the time that the nails are cut the bees are ready to run out. Keep the bees confined in the hive until the bottom-board is ready to be taken off. As the combs are taken out brush the bees into a box,—a Langstroth hive cap will answer the purpose equally well. If tobacco smoke is used in this operation, the bees will remain quiet for a long time. If they are disposed to run up the sides of the box or fly, blow more smoke upon them.[7]

After the combs are removed from the hive and the bees brushed from them, lay them flat on pieces of board. If they are crooked or wavy press them

---

[7] There are many operations which can be performed about the apiary by this means, which cannot be done without. Even the *filthy weed* has its uses as well as many things equally obnoxious and questionable in their effects on the human system. My tobacco smoker, or pipe, works quite handily here, as being held in the mouth both hands are at liberty and but slight inconvenience is experienced in handling it.

down smooth, whether there is brood in them or not, and then place another board on them to keep them flat. This should be done while the combs are warm, as when the combs are cut into the frames they will stay in place better. Cut them into the frames as soon as possible after they have been removed from the hive, placing them in as they will fit best, regardless of the position that they held in the hive, being careful not to cut them too small or so that they will fit but loosely in the frames.

I use no stick or wires to hold the combs in the frames, cotton twine (similar to that used by grocers) being better. There is seldom enough suitable brood in a box-hive to fill more than five frames. These and two or more empty combs should be placed in the hive, the bees turned in, the honey-board, quilt, or duck-cover placed on, and the hive put on the old stand, or the one which it is to occupy. The bees remaining in the room should be treated as before described. In the course of twenty-four hours the combs will be securely fastened, and the bees will gnaw off the strings and bring them out. I would watch and remove them as soon as they appear at the entrance. In a day or two, more frames or combs should be given them. If the colony is a large one put in frames prepared with starters, or those filled with foundation, alternating them between full combs. Frames of foundation (*properly fastened*) may be put in when transferred if the colony is a large one. I take the following precaution in order to have all transferred comb straight : after the strings have been

put around to keep it in the frames, lay it upon a flat
board letting the top-bar project over the outer edge,
then place another perfectly straight piece of board
on the comb and put your whole weight upon it. This
makes it perfectly straight within the frame. No
damage will be done provided the vertical and bottom
pieces of the frame are seven-eighths of an inch wide.
As the comb is just the same thickness, it will not be
injured in the least. I have practised the above for
twenty-four years. I consider good, straight, natural
comb preferable to that built from artificial foundation,
and yet strongly recommend the general use of comb
foundation.

WHEN TO TRANSFER.

Bees may be transferred at any time during the
year although it would be unwise to do so at all times.
I consider spring the best time, just before the flowers
begin to yield a good flow of honey, when the bees
have the least stores, which here in New England
is about the first of May. The condition of the bees
and season must govern in other localities. It is a
more difficult and sticky operation and more bees
will be lost, when the combs are filled with new honey,
than when there is but little honey in the hive. I
generally select either a cloudy or a rainy day, as
the bees will all be at home and we shall disturb
them the least. The time required to transfer a
colony varies according to the condition of the
combs.

UNITING BEES.

This is a frequent and necessary operation, in large apiaries at least, and an undertaking not always attended with good results, and quite frequently a hazardous one, especially to the novice. I find the following method a good one : prepare the bees by keeping them queenless three days before uniting, confining them in the hives, providing sufficient ventilation, using the screen, fig. 12, page 34, for this purpose. It would be well to put them in a cool, dark room or cellar during this time as they will remain much more quiet. If there is no brood in the hives the bees may be put in the swarming box together, after the queens have been removed. They will need feeding while thus confined. Use the glass-jar feeder for this purpose, placing it on the wire-cloth cover. The syrup should be thin, as bees while thus confined need a large quantity of water. If the swarming box is used, take the bees three days later, turn them down in front of the hive where they are to be located, uncage the queen and let her run in with the bees. If left in the hives, let us suppose (for convenience' sake) that number one is to be broken up and united with number two. First, place an empty hive on the stand which number two formerly occupied, next give the bees in both a little tobacco smoke, drumming on the hives occasionally for ten minutes to cause them to fill with honey. This done, remove the combs from number two, shake or brush them down in front of the empty hive, and then place the combs contain-

ing brood in also ; now treat number one in the same
manner, allowing the bees from both to run in together, thus thoroughly mixing them, letting the queen
run in with the bees as in the former instance. Bees
can be united without risk by the above plan. I find it
difficult to unite blacks and Italians in any other way.
Bees seem to recognize each other by scent. Hence,
when two colonies have been thoroughly fumigated
with the same aromatic material, they can be easily
united.

I find it a good plan to feed bees with sugar syrup
scented with peppermint, or something of the kind, for
several days before uniting them. They will thus
become scented nearly alike as they have partaken of
the same food. These bees should be thoroughly
shaken up and their honey sacs well filled when united.
This tends toward keeping them good-natured during
the operation. Well-filled stomachs have a tendency
to soothe their combativeness. This same cause seems
to produce a like effect upon human nature.

THE UTILITY OF UNITING BEES.

There is nothing gained by uniting two large colonies in the fall, and the operation would prove
worse than useless. If there are a number of weak
ones in the apiary it is advisable to unite them by
putting two or three in one hive, in all cases retaining the best queen for the colony. When two large
colonies are united, double the stores will be con-
sumed, and when the bees are ready to commence

work in the apiary such a colony will contain only about the same quantity of bees it would have had, had none been united with them, as bees that are united die off rapidly in the spring; and if one colony has the bees of two queens they will diminish on this account twice as rapidly as they will in one that has only the bees from one queen. There is, therefore, a disadvantage in uniting two large colonies in the fall.

UNITING BEES IN SWARMING TIME.

There is much to be gained by uniting bees in the swarming season. This would not be advisable until the apiarist has as many colonies as he wishes to keep. When two swarms are to be thus united, the sections should be placed on the hive immediately, and the result will be highly favorable if there is a good honey flow at the time. Swarms thus united seldom quarrel; the extra queens will be disposed of and business will progress favorably. I would recommend the above plan when the apiarist has all the colonies that he wishes to care for.

WHEN AND HOW TO REMOVE CELLS FROM A COLONY THAT HAS JUST CAST A SWARM.

In case the apiarist is not particular regarding the purity of his stock, he can obtain queens to fill the places of old or infirm ones by the following plan:

during the swarming season, prepare nuclei and save the best cells built in any of your best working colonies, by cutting them out after the swarm has issued.

The proper time for doing this depends largely on the time when the bees swarm. If the weather is pleasant for two days preceding this, the cells may be safely cut out the seventh day after; but if unfavorable for several days previous, it is quite difficult to determine exactly when the cells are sufficiently matured to be removed, as the colony may have been prepared to cast a swarm several days before they did so and were prevented on account of unfavorable weather. Or, possibly, they may have made every preparation to swarm during the time they were confined to their hives by stormy weather, as they often do. In such case, the colony should be examined and one or more cells opened near the base, for the purpose of ascertaining as near as possible the age of the embryo queens; and if not sufficiently matured to be safely removed let them remain a few days longer.

In a day or two after the colony has cast a swarm prepare nuclei, and by the sixth day cut out the cells leaving one of the best for the parent stock.

In large apiaries a plenty of cells may be obtained in this way, and all the old or infirm queens may be superseded at any convenient time. I would recommend for this purpose the small hives, fig. 1, page 4.

If the apiarist has a large number of colonies and wishes to keep his stock pure, he should remove his fertilizing nuclei half a mile or more from other bees and place a colony containing drones of such strain as he wishes to propagate, in the same yard with them. Not one queen in twenty-five will be impurely mated, notwithstanding there may be hundreds of colonies of impure bees within one mile of them.

THE PROPER AGE TO SUPERSEDE QUEENS.

In apiculture, the same as in stock-raising, the poultry-business, and like industries, the age which completes usefulness depends largely on the amount of labor performed, and we find that the rule "that all animated beings outlive their usefulness" applies equally well to queens. Hence I consider it a good idea to supersede them as often as once in two years. A good, prolific queen will lay nearly her full quota of eggs during this time, especially under the present system of managing bees, and will certainly have spent her best days during that period. It will be advantageous to keep some queens longer, but this is the exception and not the rule. Young and vigorous queens are the profitable ones, and I do not care to keep old queens for any purpose. While some claim that the breeding queens should be two years old, I prefer those that are young and vigorous for reasons previously stated. When a queen is in her second year she is in her prime as queen mother. Again, the cost of queens is so reasonable, that the expense of often requeening a colony is but trifling when compared with the advantage gained.

CHAPTER XI.

SPRING AND FALL FEEDING.

Here in New England, and in the northern portion of our country generally, we usually experience cold and windy weather during early spring; warm spring weather begins late, the portion of the season during which the surplus honey is gathered, being of short duration, and the fall, during which we must prepare for winter, terminates only too soon, as early frosts cut off the late honey forage. From this we draw the following conclusions: first, unless colonies are strong in numbers when the honey flow comes, a large percentage will be lost; secondly, the long dearth of honey from the expiration of the last flow until winter's cold and chilling blasts sweep over the land, and the hills and valleys are robed in white and our pets begin their long rest, must be counteracted by artificial means. The former can only be accomplished by

SPRING FEEDING.

The reader is well aware that bees will start out in search of water and pollen as soon as the snow begins to melt and disappear, and the first flowers bathe their petals in the warm sunshine of early spring, often being obliged to travel a long distance for both, the results being, that thousands are lost on their way home; in fact, it is not unusual to see the ground literally strewn with dead bees, exhausted by their long search and struggles against the cold winds, perishing almost within reach of their homes.

In order to counteract this begin building up the stocks by early breeding, stimulating the bees[8] by feeding with thin syrup and artificial pollen; and when the first honey flow comes, the colonies will be literally boiling over with bees ready and willing to take advantage of it. The above is especially important to queen-breeders as they must have a good supply of early drones to fertilize early queens. Bees need large quantities of water and pollen; the former may be supplied in the shape of the thin syrup previously mentioned, and I would recommend for this purpose Locke's feeder, described at end of volume; the latter by placing boxes containing wheat flour[9] (with a piece of comb placed in it upon which they may alight) in some warm, sheltered portion of the apiary. Reason and experience have taught me that they will accept and make good use of this substitute for the natural pollen to the extent that they work on it, flying to and fro between the box and hives in perfect clouds, filling the air with the music of their busy hum and continuing until the natural yield is abundant. It is quite pleasing and

---

[8] For spring stimulating, the food should be placed directly over and in immediate connection with the cluster, so that it is accessible to the bees in the coolest weather.

[9] I have found that wheat flour is the best substitute for natural pollen, as the bees can carry a larger amount of it into the hive in a given time than of any other.

interesting to watch them when at work on the wheat flour, tumbling over each other in their hurried efforts to make the most of their time, and rolling up and packing their loads on their legs.

THE AMOUNT OF FOOD TO GIVE EACH DAY.

In no case would I give any colony over one-half pound of syrup unless they were short of stores. In such cases give two or three pounds to commence with as a cold spell might set in and the bees starve rather than leave the brood to obtain food. Feeding should be discontinued just as soon as they begin to gather honey from the flowers. I generally give them all the wheat flour they will carry in. They will stop working on it when the natural pollen becomes abundant. Colonies properly stimulated will be certainly ten days in advance of those allowed to build up in the natural way. The fall honey dearth (previously referred to) may be counteracted by

FALL FEEDING.

It is not generally necessary to resort to feeding in the fall, except for the purpose of supplying the bees with necessary winter supplies.

For this purpose an entirely different quality of food and method of feeding must be adopted. Late feeding sometimes pays, but generally those fed later than September 20 (here in the north) will not winter well; hence, all the necessary supply should be given before that date and sealed over before cold weather sets in and drives the bees from the outer to the centre combs. If not, the syrup will not be sealed before spring and will sour and ferment, running out of the cells and besmearing the combs, and when spring opens they are mouldy and damp. If there is one thing more than another that the bees dislike, it is to have the combs in the above condition, and if obliged to eat the sour syrup before spring, it is almost certain death to them, unless the weather is so favorable that they can fly out often. I would recommend early fall feeding when it is not advisable to unite the colonies.

FOUL BROOD.

It is exceedingly fortunate that the honey-bee is subject to few diseases, else the business of apiculture might be very precarious; but as it is, a person can begin the occupation of bee-keeping with the assurance that a good hive and locality, together with careful management, will prove, to say the least, as sure and remunerative as any other occupation with the same investment of capital.

There is one dread scourge, however, that at times infests an apiary, whose ravages are so fearful that a short article may well be written in regard to it. I refer to foul brood.

The origin of this disease is as yet unknown, and

its history obscure; many theories in regard to it
have been written, none of which however have as
yet been proved, and all seem to exist in the im-
agination of their authors.

I propose to give a description of this disease, the
means that have heretofore been used to attempt a
cure, and the plan I should adopt to eradicate it
should it show itself in my own apiary.

Foul brood is the most fatal to bees of all other
diseases. It is a disease of the larvæ only; the
sealed brood die in the cells producing a most intol-
erable stench, which of itself would be sufficient to
one who had ever seen a case to recognize it again,
as it may be perceived at some distance from the hive.

HOW TO DETECT FOUL BROOD.

Upon opening a hive infected with this disease the
cappings of the cells are found to be of a dark,
yellowish brown, depressed in the centre, and usually
with a small hole the size of the point of a pin in the
centre of the caps. Upon opening the cells the brood
is found dead, partially decayed, moist and slimy
in form, and emitting a noisome stench such as no
other cause can produce. On its first appearance a
few cells only may be found affected; but if allowed
to go on without anything being done, in a short time
every cell of brood will be found contaminated, and
ere long the colony will die out for want of young
bees to replenish it; and not only this, but unless
active means are at once taken, every colony within

flight range will be certainly and surely affected also. No half-way measures should be taken, but active means adopted if one desires to be rid of this terrible pestilence. As this scourge affects the larvæ only, the first thing to be done is to remove the queen, for when we stop the production of brood, we so far check the further growth of the disease.

Attempts to cure this disease by means of disinfectants have utterly failed. The hive, frames and comb have been thoroughly sprayed and washed with chlorinated soda and other purifying agents with no avail, and the disease was until recently pronounced positively incurable. The recent discovery of salicylic acid, it is now claimed, has produced a means of curing this disease and eradicating it from a hive. I have never seen it tried and have little faith in it as a means of cure, especially in the hands of the inexperienced, and do not recommend its use. One thing is fortunate, the queen may be used again as it has been fully proved that she does not carry the disease.

If my apiary were infected with this disease, I should remove the queen from every infected colony, and at once utterly and totally burn and destroy every trace and vestige of hive, frame and comb, even going so far as to burn up the stand upon which the hive containing the infected colony was placed.

The disease is known to be of fungous growth, and the infinitesimal spores are carried from one hive to another on the legs or bodies of the bees, or may be conveyed into a hive by means of feeding contaminated honey; hence the necessity of utterly destroying everything concerning which a suspicion of contamination exists.

If any one desires to attempt a cure of *this scourge*, he will find in a little book written by Mr. Muth, entitled "Helps and Hints," his plan of cure and the means which he adopts, by the use of salicylic acid; but for myself I do not think it will prove a paying experiment, for I should fear to find the disease breaking out anew season after season, unless I totally eradicated it in the beginning.

FIRE, THE BEST MEANS OF CURE.

Fire will certainly destroy every vestige of the pestilence, and is the only known means as yet of so destroying it. I would therefore advise all, who find their hives infected, not to delay an instant, but at once proceed to make a bonfire, and be sure that everything connected with the apiary which has been exposed is totally burned. The virulence of this disease is such that it can be carried from one colony to another upon the hands of the operator; so one must be exceedingly careful to cleanse thoroughly the hands and all tools used upon a suspected hive, before operating upon one known to be disinfected.

If one thinks my plan is too severe, he certainly has my permission to use any other that he chooses, but may rest assured that by so doing he will find plenty of time in which to repent, and I fear will wish he had repented first.

The apiarist should keep a watchful eye and ever be on the alert for indications of this disease, when examining the combs of his colonies. I seldom open a hive for any purpose without carefully examining each comb of brood to see if there is any trace of this dread disease. Although it is sixteen years since it existed in my apiary, I have not forgotten my experience in that line.

If even one infected cell is found, the colony in which it appears should be isolated three miles at least from the other apiaries. If only a few cells are infected I would cut them out, taking with them one square inch of the adjoining comb, and burn them immediately.

In adopting my remedy it will not be necessary to destroy the combs and honey. The hive should be consigned to ashes, but the combs and honey may be preserved; only great care must be taken that the bees do not have access to it, as the disease would spread rapidly and every colony in the yard would soon become infected. The honey may be " strained out," and clarified by heating for several hours in a vessel placed in water and the combs melted into wax. Do not then give it back to the bees. There might not be any danger of communicating the disease by so doing, but with my experience the risk is considerable, and no apiarist can afford to trifle with a disease so contagious and so devastating.

There may be some remedy that will temporarily check its ravages, but as a positive cure I cannot

6

have any faith in it whatever. If you have this disease in your apiary destroy every infected colony at once, even though they all have to be sacrificed in order to rid the bee-yard of it.

CHAPTER XII.

WINTERING BEES.

Probably no subject has ever been presented to the bee-keeping fraternity for consideration, of more weighty importance, or upon which success in apiculture more largely depends than the above. The apiarist is often favored with long and apparently exhaustive essays upon it, and it has been thoroughly discussed both through our journals and at our conventions; yet it remains, apparently, an unsettled matter, and we still hear reports every spring, from different parts of the country, of heavy losses in wintering. I do not claim to have solved this problem, but merely intend to add my opinion and experience to what has already been said. It is generally conceded by a large number of our best bee masters that the proper place to winter bees is on the summer stands, and in this I agree with them. This method is being generally adopted. In order that the bees may winter well on summer stands, they must be properly prepared in the fall. Double-walled hives are a necessity in our northern climate; and I do not consider chaff-packing of any consequence whatever, a dead-air space between the outer and inner walls being all that is needed. A colony should have twenty-five pounds or more of honey (mostly sealed) to winter on. Be sure that they have young and vigorous queens in the fall, and are strong and populous the first of October.

I do not consider upward ventilation necessary when bees are prepared as follows: place two sticks crosswise over the tops of the frames about four inches from the ends, two more lengthwise five inches apart, resting them on the former, cover this with some bagging or other coarse cloth and place the chaff cushion over it. Early in October, bore a one-inch auger hole three inches down from the top and halfway between the front and back of the hive, take a stick one-half inch square and twenty-four inches long, pointed at one end, and slowly twist it through the combs to the opposite side of the brood-nest, thus making a clean winter passage for the bees. Very few bees will be killed by this operation, and in all my experience, I never have killed a queen while doing it.

Set the hive two or three feet above the ground so that it will be out of the way of the snow. I have found that where the snow was banked against single-walled hives the bees were nearly all dead, the frost, cutting in through the snow and ice and penetrating several inches into them. A board should be placed in front of the hive to shield it from the sun. Now if the bees can have a flight during the month of February they will generally winter successfully. A few bees will be lost in the snow but they may as well die there as in the hive, but they should not be permitted to fly unless the glass stands at 50° in the shade, and the weather very calm. If they cannot fly once during the winter they will not come through well. All the theories about dry fœces and other nonsensical talk will pass for what it is worth. Plenty of sealed honey, thorough packing, and a good fly once or twice during the winter, are the plain common-sense requirements which insure us good, strong colonies in the spring. Lacking these, we shall find ourselves with many empty hives and perhaps many weak colonies when spring opens.

SUGGESTIONS ON CELLAR WINTERING.

As my ideas of cellar wintering differ from those of most bee-keepers, I will describe my method. I think that where bees are wintered in cellars, spring dwindling may be prevented by the following plan: leave them on the summer stands until about December 1, then carry them into the cellar and close it. This will, of course, warm them up. In the course of ten days I would open the cellar some cold night and run the temperature down to about 25° above zero, allowing it to remain so about twelve hours or until the bees are thoroughly cooled off. I would then raise the temperature to about 35° or 40° keeping it at that point for a week or 10 days, then gradually lower it again, and so on through the winter. This would give the cellar a temperature that would conform to that out-of-doors, but the advantage would be in not having such extreme low temperature in the cellar. The long-continued cold spells both in winter and spring are one great cause of loss in wintering and spring dwindling. Colonies that are put in the cellar in November before the cold weather sets in, and kept there until late in the spring, are those which dwindle after being placed upon the summer stands.

The bees are often permanently removed from the the cellar in April, and in our northern climate especially, the temperature frequently runs as low as 10° above zero during this month. I have seen it stand at 28° above as late as the tenth of May. This is very unfavorable for bees that were wintered in cellars or repositories.

SHOULD THE HIVES FACE NORTH OR SOUTH?

It is the usual custom and practice to face the hives toward the south, and while there are some good reasons why they should face this way, yet there are equally as good reasons why they should not.

During the hard winters of 1879 and 1880 a large portion of the bees in the northern states perished on account of the severity of the weather, and long confinement to the hives with no opportunity for a cleansing flight.

I have a bee-keeping friend who had twelve good colonies the fall previous, which were in a bee-house, six of them facing the south and six the north, and all those facing the south died while those facing the north came through all right, and in fine condition. Now what is the philosophy of this?

My opinion is that although the bitter cold weather confined the bees to the hives through many long months, yet the warm sun, striking on the south side of the building, warmed up the fronts of the hives facing that way, inducing the bees to fly at unfavorable times, few of which ever returned, and in this way the colonies were gradually reduced in numbers until the few remaining bees perished.

On the contrary, the bees facing the north were not thus enticed to leave the hives, but remained quiet until the weather was warm enough to insure a good cleansing flight and a safe return; consequently they were strong in numbers and came out well in the spring.

Hives that face the south should be kept well shaded from the sun; boards may be placed against the front of them for this purpose.

When bees are confined to the hives for several months and are about to die for want of a cleansing flight; the least disturbance will cause the loss of large numbers of them, hence they should be kept perfectly quiet until some warm, pleasant day when the boards should be removed from the fronts and the bees have a good flight.

Colonies that stand in warm, sheltered locations, facing south, will fly out as the sunshine warms up the front of the hive slightly, even when the glass indicates a temperature several degrees below the freezing point, while, if the hives faced the north, they would not attempt to fly.

DOES LIKE PRODUCE LIKE?

Those who purchase *dollar queens* intending to use them for breeding queens have been badly disappointed in many instances. They expected that such queens would produce *all* three-banded workers, and that all the young queens would be duplicates of the mother. All queens *should* be pure as far as mating is concerned, as no queen-breeder, who is a master of apiculture and who means to deal honestly, will permit drones of several races in or near any one of his fertilizing apiaries, as this *can* and *should* be prevented; nevertheless, hundreds of untested queens may be sent out which will be pronounced impurely mated, when in reality they have been *purely* mated.

It is a well-known fact among stock-breeders that the progeny of prize stock will not always show all the markings and qualities of the original: for instance, when breeding from imported stock of the famous Jersey herd of cattle, the young cows quite often fail to prove as handsomely formed, as clearly outlined or as good milkers and butter producers as their mothers, and yet they are pure and valuable. Again, with horses, the dam may be perfect in every respect, and the sire of the choicest, and both have a 2 : 22 record as trotters, and yet it is only occasionally that their colts prove as valuable or make as good a record as their parents. These comparisons may be indefinitely extended.

This law holds good throughout nature, and applies equally as well to queen-rearing; and the apiarist, who thinks that he can enter it without a thorough knowledge of it, and produce first-class queens and those that will duplicate themselves and, at the same time, compete successfully with experienced breeders, mistakes his calling.

It is only by the best selection of stock and most careful breeding, that queens will be reared which will produce daughters as perfect in all respects as themselves, and no queen-breeder will part with such a queen for a small sum.

When a customer purchases an untested queen, he takes his chances regarding her value; if she prove equal to the mother he has obtained a valuable queen; but if, on the contrary, her progeny is poorly marked (and for this reason pronounced hybrid), it is simply a freak of nature and not a concocted plan of the dealer to cheat his customer.

No breeder can possibly determine how valuable the progeny of his untested queens will prove; nevertheless, if he understands his business he *will* know that *every* precaution has been taken in breeding and keeping them pure, and nature must decide the rest.

I have had customers write me that the queens received were impure and that I must have black drones in the apiary; I must confess that such reports are rather discouraging.

I well remember the history of the first Italian queen that I purchased of Mr. Parsons twenty-three years ago ; out of twenty-six queens reared in the first lot, only two were duplicates of the mother. Well, the queen purchased of Mr. Parsons was reared for him by one of the most experienced and honest apiarists in the country, but it was no fault of his ; the trouble was with the imported queen sent him. And this is a fair sample of all imported queens.

It was many years after this before queens could be produced or procured which would duplicate themselves, and I never had any until I produced them myself by careful selection and more careful breeding.

Before leaving this subject, I wish to state that every apiarist wishing to procure the best stock should forever banish the idea that he will find it in the shape of cheap queens, bred by the novice or in a careless manner, and also that the tested queen is the only proper one to purchase from which to breed.

PARTHENOGENESIS.

The theory of parthenogenesis, as applied to the honey-bee, viz. : that the eggs of a virgin queen will hatch, and that the product thereof will invariably be drones, was first taught by Dzierzon some fifty years ago, and although for awhile it had many opponents, it has now become a well admitted and established fact in bee-culture, and he who disputes it shows that he is either wilfully or ignorantly obstinate.

Were the honey-bee the only example of this peculiar state of things existing in nature, doubts in regard to its truth would be more excusable, and a much greater cause for wonder and conjecture; but when we find that many others of the inferior order of creation are subject to the same law, we have no reason to doubt in regard to our bees. Why this should be so, we do not understand, but the all-wise Creator had some wise end in view, presumably, that when a worn-out queen is superseded at a time when no drones exist in the hive, the virgin queen thus produced may be able to raise drones for her own fertilization, in order that the colony may not become extinct as it otherwise would.

Another theory has been advanced in this connection: "that the pure queen, however mated, must produce a pure drone of her own variety." This theory I cannot accept. It is well known among breeders of horses, cattle, etc., that if a female of one breed mates with a male of another, that such female never again produces her like in absolute purity. So with our domestic fowls. If a pure white leghorn pullet mates but once with a black Spanish cock, her chicks ever afterwards are liable to show an occasional black feather. With the mammals it may be said such impurity is caused by carrying the young in the uterus, and thus the blood becomes impregnated with the blood of the sire; but with the fowl, such cannot be the case, and the impurity must be caused by absorption of the male semen. If such is the case with fowl, why is it not so with the queen bee? This I know is only reasoning from analogy, but the presumption is that such reasoning is correct; at least it shows it to be possible if not highly probable, and in the matter of the purity of our queens we want no question of possibilities even.

I advise, therefore, all who wish to keep their stock absolutely pure, to allow no drones from hybridized queens any chance to mate with such queens as they desire to rear to breed from, if they expect to keep their stocks absolutely pure; I would as soon send, as purely mated, a known hybridized queen, as one mated with such a drone.

CHAPTER XIII.

KEEPING BEES FOR PLEASURE AND PROFIT.

There are about eighty thousand bee-keepers now in the United States; were there twenty times as many engaged in that occupation, there would still be room and forage for as many more. Thousands of tons of honey go to waste every year simply because there are no bees kept to gather and store it.

Hundreds are prevented from taking up this occupation by dread of the sting of the bee, who otherwise would gladly enter into it. I cannot say that it was not the intention of "Him who created all things," to prevent too large a propagation of this insect, by giving it so powerful a weapon, but it was given for some wise purpose and this is perhaps that purpose.

It is well known that thousands of insects visit flowers, both day and night, in search of honey and pollen, and that this same honey and pollen are secreted in the flowers to attract these insects thither. For what reason are these insects thus attracted? Simply that they may carry the fertilizing principle from one flower to another, and thus cause a yield of grains and fruits, where otherwise sterility would be the result; and were it not for this secretion of nectar the bees would not be attracted to the flowers, and consequently this fertilizing would not be performed.

In seasons when the flowers are abundant, only a small part of the honey is gathered from them that might be, were the bees on hand to collect it.

The flow of honey depends much upon the state of the atmosphere; far more so than it does upon the quantity of flowers in the fields and woods. I have seen the fields white with clover blossoms, and still found no honey being secreted in them, on account of the state of the weather, and have seen bees starving when the trees were in full bloom, but the weather was such that they could not leave the hives in search of forage; when, by a simple change of the wind from east to west, there would be more honey secreted than the bees could by any possibility collect.

Here, in New England, near the seacoast, we are troubled with east and northeast winds nearly all the time during May and June; this is unfortunate for us, as the bees can do but little at such times; notwithstanding this fact, large crops of honey have been gathered here. About four years in five, bees will, in any locality, pay a handsome profit on the capital invested. The weather in all parts of the country is a serious drawback to bee-keepers, and is about the only thing that the ingenuity of man cannot control so far as apiculture is concerned.

Thousands of people are engaged in bee-keeping in the United States, who give their whole time and attention to the business, while there can be found in most every township a number who keep a few bees for pleasure, and to supply their tables with pure honey.

WHO SHOULD KEEP BEES.

People who are incapacitated from hard labor may keep bees, for while there is much hard work to be done in a large apiary, it does not compare with the labor required in farming. The most that is required in bee-keeping is to do what is needed just *when* it is needed, and not put off for a moment the thing that is required to be done.

Men well advanced in years can adopt bee-keeping as a means of employment, with much pleasure as well as profit. The business can also be well adopted as an auxiliary to some other light employment, and a profit made in both, where either alone would not prove sufficiently remunerative.

When only a few bees are kept, the time required to care for them would take but a small portion of the day, except during the swarming season and the honey harvest.

Professional men, such as clergymen, doctors and lawyers can keep bees, with pleasure and profit, and thus obtain a large amount of recreative exercise; mechanics, although seldom at home, can keep a few colonies, as they can do all the work required, before and after their daily toil, and thus add a few dollars to their revenue, and at the same time, profitably and pleasantly employ their leisure moments.

Ladies can keep bees as well as men; in fact, there are hundreds of ladies now engaged in bee-culture, who find it a healthful and remunerative occupation. They soon learn to handle bees, and there is nothing about the business but what is perfectly proper for a lady to attend to or engage in.

No one should engage in bee-keeping, however, who has not made up his mind to read carefully, study hard, and think deeply. Success may for a time follow the ignorant or careless, but ere long failure will be the inevitable result. Those of our bee-keepers who have studied and labored the hardest, and given the most thought to the subject, have invariably been the most successful. Not that it requires much time or labor to care for a few colonies, as any man who has a love for the business (and none others should attempt it) will find plenty of time to be spared from his vocation or profession to care for a half dozen colonies, and find a pleasure and enjoyment therein, which, even if no pecuniary gain were made, will amply repay him. He who studies in the wondrous realms of nature will always receive a vast amount of substantial advantage. And he who engages in apiculture finds that not only does the field of entomology open itself to him, but horticulture, floriculture, arboriculture,— in fact, the whole extent of the botany of nature is only secondary to the pursuit; as he who becomes acquainted with the flora of his locality, and the seasons when the blossoms expand and the flowers yield their nectar, will be the better able understandingly to care for his colonies, and better know when to aid them by judicious feeding, or to remove for his own use the honeyed sweets the busy bees so persistently gather. The whole tendency, too, of apiculture, is to expand the mind and enlarge the understanding; taste must be refined and thought elevated by association with the most wonderful of God's creation among the inferior classes.

HOW TO COMMENCE BEE-KEEPING.

A word of advice on this point to beginners may be of value to many. If I were intending to start anew in bee-keeping, I would purchase one colony of bees from some responsible dealer and get it in a movable-frame-hive (the Langstroth principle I consider the best) ; this would give a good start and save all trouble of transferring, which would have to be done if the colony were bought in a box-hive. I would purchase some standard works on bee-keeping, say, "Langstroth's on the Hive and Honey-bee,"

Quinby's "New Bee-keeping," or "Cook's Manual,"
and also subscribe for some good bee journal.
Thus I would be armed with the necessary literature
on bee-keeping, and ready to study up the theory, as
well as to attend to the practice of bee-culture. I
would devote all my leisure moments to the study of
these works, in order to get posted, as soon as possible,
on the habits of the honey-bee, and the best methods
of managing an apiary. Books are of little value,
unless one, at the same time, practises the ideas gained
from them; this can be done by opening the hive and
examining the bees as often as desired; thus, practice
in manipulating a hive, as well as courage in handling
bees, will be gained. Of course, a hive should not
be often opened while the bees are bringing in stores,
not that the bees are injured thereby, but the disturbance thus caused would make the colony less profitable,
as it would prevent some of the bees from bringing
in as large an amount of honey as they would were
they undisturbed.

The best time to open and manipulate a hive for
experimental purposes is just before sunset.

BEST LOCATION FOR AN APIARY.

Having advanced so far that one can handle his
bees easily and boldly, and having also obtained, by
study and observation, a fair insight into their habits,
I would extend the business somewhat by procuring
more colonies. If I were bound to keep bees, and my

present location was a poor one, I would keep less colonies, or move to a locality which offered better prospects, and where forage was abundant. Most any part of New York state will afford a good locality; but in order easily to market one's products, and obtain cheap freights on supplies, the line of some leading railroad should be chosen as a spot on which to locate permanently.

REQUISITE IMPLEMENTS FOR THE APIARY.

One of the first tools to be procured, and the one that is most required in managing bees, is a good Bellows smoker. There are a number of good ones from which one may choose.

A good extractor is a very necessary article in an apiary; there are many excellent ones made, from which I am hardly able to make any choice as all are good. I should, in purchasing one, procure it of the supply dealer nearest to me in order to save on expressage or freight. With these tools, and a supplement of hives and frames, one is ready to commence bee-culture, but unless a full determination is made to devote time and energy, mind and hand to the business, one had better let it alone. But if a person takes hold of it with his mind fully made up to devote all his energies to it, and to be ready at all times to take advantage of every point which tends to help the business along; not to be easily frightened or discouraged at little reverses, should they happen, he will make a success of the operation, and make a larger per cent of gain upon the capital invested, than in any other occupation.

Not only does it yield a handsome profit, but it brings to our table a delicious article of food such as the gods themselves might relish and desire. Health demands that we should use some sweets, and what one more palatable than pure honey is known to be? for, produced from our own hives, with the knowledge that it is the result of our own care and labor, it will become doubly delicious and attractive.

MANAGEMENT OF THE APIARY;

OR, THE

PRODUCTION AND MARKETING OF HONEY.

BY GEORGE W. HOUSE.

MANY points must be taken into consideration in giving instructions on this subject; in fact, a whole volume might be written without going into the little details that are necessarily connected with the manipulations, many of which are unexpected and unavoidable, in obtaining the largest possible amount of surplus.

To secure all the surplus honey obtainable, the apiarist must have the ability to meet readily any emergency, however unexpectedly presented.

Manipulations that would prove the most successful one season, if put in practice at the same time of the following season, and in the same manner, would be most disastrous to our colonies. Therefore we must study the seasons and prepare ourselves to manage our apiaries in accordance therewith. All the advice I can give as to the proper time of having our bees in the desired conditions to store the coming nectar advantageously is, that we must have a knowledge of the time and extent of the bloom about to make its appearance, always keeping an eye on the weather.

(99)

THE HIVE.

The hive is an important factor in *all* our manipulations. It matters not so much about the style of frame used, whether Langstroth or Gallup, Quinby or American; but the hives should all be alike, and constructed as simple as possible, avoiding all loose pieces, etc. All hives *should* be two-storied, both bodies exactly alike and interchangeable, so that we may run our apiaries for either extracted or comb honey, or both, changing at any time we wish and without any unnecessary expense. I will endeavor to explain the most important points in connection with the hive, as I proceed with the subject.

SPRING MANAGEMENT.

As soon as the weather becomes warm and settled, and the bees bring in the first pollen, the apiarist should commence preparing for the season before him. At this season of the year the bee-keeper will find his colonies varying in strength from a two-framed nuclei to the colony that is overflowing with bees. He should now aim to have his colonies brooding to their utmost capacity, or all that the bees can properly care for. This is accomplished in different ways; many go through their apiaries, select the weakest colonies and double them up by uniting two and three together; but, after repeated trials, this method has been abandoned by the skilled apiarist, it being found to be impracticable. After repeated experiments the following mode is considered the most practical and the one giving the best and quickest results.

CONTRACT THE BROOD-CHAMBER.

Go through the apiary and confine each colony (by means of close-fitting division boards) to as few frames of comb as they can properly cover and care for. While the bee-keeper is thus going through the apiary he will notice that the outside combs containing brood will, in most instances, have brood on but one side of the comb, and *that* on the side of the comb toward the cluster of bees.

These frames should be reversed, putting the side containing but little or no brood inside, or towards the centre of the hive. The reader will at once see the advantage gained. By contracting the brood-chamber the bees are enabled to keep up a higher temperature, and the combs being reversed the queen finds more room within the cluster to deposit eggs, thus increasing the size of the brood-nest and facilitating the increasing strength of the colony.

EXCHANGING COMBS.

We now go to our strongest colonies and take from each one a frame of hatching-brood or brood about to hatch, putting in its place a frame of honey or empty comb. Give one of these frames of brood to each of the weakest colonies by moving the division board, spreading the brood-frames and inserting the frame in the centre.

We should repeat the same treatment every week or ten days until our colonies are all of equal strength, which should be accomplished by about the first of June, or the time when white clover begins to bloom. The bee-keeper, who has never practised the above method, would, upon adopting it, be most happily surprised at the rapidly increasing strength of his weakest colonies, and all attained *without injury*, but *beneficial* to his strongest colonies. This building-up method is *the key* to our success for the season before us.

FEEDING FOR STIMULATING.

A judicious stimulating, by feeding wheat, rye or oat flour, and honey or sugar syrup, will pay the apiarist well for his time and money spent. The flour should be placed in troughs or boxes, and put in some warm place in the apiary that is protected from the winds, feeding only on the warm sunshiny days of early spring. As soon as the bees get natural pollen, it will be unnecessary to feed flour, but we should change to honey or sugar syrup. When once commenced, the feeding should be kept up until the white clover bloom, discontinuing it only at such intervals as when the bees are bringing in honey enough to excite the queen to laying, as during the willow, fruit tree and the raspberry bloom in this locality.

The management described brings us to the time when our principal honey-producing plants begin to bloom, and the swarming season about to commence. We should now have our queen-rearing nuclei, or colonies for that purpose, in operation, so as to be able to keep a laying queen in each hive. The manner of rearing and introducing the queen I will not attempt to describe, but refer the reader to other pages of this book, whose author is acknowledged the highest authority, and who is not excelled by any one in the world.

We are now ready for the busiest part of the season, and we will suppose the apiarist is working for comb honey. The first thing to be done is putting on the surplus arrangements, and I shall endeavor to show the reader the usefulness of a two-storied hive.

Much depends upon the construction of the surplus arrangement in facilitating the various manipulations and saving much labor in preparing our comb honey for market, in the way of keeping the sections (as far as possible) free from propolis, wax, etc. For this purpose nothing is better than the so-called *broad-frames*.

If your hives contain frames of the Gallup or American style, the broad-frames should each hold four section boxes $5\frac{1}{4} \times 5\frac{1}{4} \times 2$, known as the two-pound box. The hive should be constructed to take eleven brood-frames, but in place of three of the brood-frames we use two broad-frames, one on each side of the brood-nest, thus the brood-chamber contains but eight frames, which is a sufficient number from which to obtain the best results in that style of frame.

The second or upper story should be of the same dimensions as the brood-chamber or lower story, and will hold eight of the broad-frames, with thirty-two sections, making forty boxes in all holding eighty pounds of honey to each hive. If the apiarist is using the Langstroth hive or frame, I would advise using broad-frames holding eight of the $4\frac{1}{4} \times 4\frac{1}{4} \times 2$, or so-called pound sections, putting one on each side of the brood-nest, which should be composed of but seven or eight frames, all that is necessary for securing the largest amount of surplus.

The upper story should be of the same dimensions, inside measure, as the lower story, and should contain seven broad-frames holding fifty-six boxes, or seventy-two all told, holding about seventy pounds of honey.

It is best to arrange the boxes at the sides of the brood-nest, a week or so before putting on the top boxes, and when the upper surplus arrangement is put in position, exchange two of the broad-frames in the centre for the two at the sides of the brood-nest. With such management, the bees commence work in the sections much more speedily.

The bee-keeper frequently has colonies that refuse to commence operations in the surplus boxes; they may be in the best condition possible, but on account of their having the swarming fever, or possessing a characteristic for crowding the brood-chamber with honey, they are loath about entering the boxes. By hanging a frame of brood in the centre of the upper story and boxing on both sides, such colonies will go to work with an almost incredible rapidity.

I might state here that at present I am experimenting with a reversible frame, which I think will be most effectual when applied to colonies that are reluctant to enter the boxes, as the brood is thus placed at the top of the comb and nearest to the surplus arrangement, with the honey at the bottom of the frames, which is contrary to the laws of the hive; the bees will at once move the honey upwards and into the boxes. The bee-keeper, however, must exercise good judgment in such a manipulation of the brood department, lest it may prove injurious to himself and the colony.

After work in the surplus boxes has been thoroughly established, the apiarist should know that every colony has just room enough, and no more than they will work advantageously.

As fast as the boxes become finished, they should be removed and empty ones put in their place. To have the sections neatly and closely filled, reverse the boxes by turning them bottom side up. For this purpose, I am using reversible broad-frames; by their use, the labor in manipulating is much lessened in more than one way.

COMB FOUNDATION

If used in the new hives will pay about 300 per cent. on its cost, at present prices. If used in wired frames it may be as light as eight square feet to the pound; but if used in frames not wired, it should measure about five feet to the pound. For surplus boxes a very light foundation is recommended by many of our most successful apiarists; some advise using it as light as fourteen feet to the pound. In this I must disagree with them. I prefer a foundation with very thin base or septum, and high, heavy side walls measuring about seven and one-half feet to a pound. For several reasons I prefer to have the sections filled with such foundation, and believe the time not far distant when a majority of our apiarists will support me in my views on this subject.

Foundation for the surplus boxes is best when newly made, the bees accepting it much more readily. The bee-keeper should manage to have such foundation fresh every week.

I think our foundation mills can be greatly improved by having the dies cut deeper, so that the septum of the foundation can be made quite thin, and giving plenty of room for the formation of the side walls, that they may not be pressed so hard, and left with so smooth an edge, but come from the mill soft and pliable, with high side walls whose edges should be ragged.[1] Such foundation is accepted by the bees and worked out in about one-half the time required on other foundation.

---

[1] Since writing the above Mr. W. C. Pelham, of Kentucky, has placed upon the market a new Comb-foundation Mill, which makes the best foundation I have ever seen. The mill is simple in construction, easily and quickly adjusted and operated, and the price so small that every bee-keeper can afford to own one.

EXTRACTED HONEY.

In working for extracted honey, the same management is required, up to the time of putting on the surplus arrangement, as with comb honey.

Instead of putting the broad-frames with section boxes in the brood-chamber at the time stated, we must spread the brood-frames and insert an empty comb. If the weather permit, this should be repeated every four or five days until the hive is filled with comb and brood. The upper story should now be filled with frames of clean empty combs, first extracting the honey from the brood-chamber or lower story, spreading the brood and returning the empty combs to the centre of the hive. No bee-keeper, however, should extract honey unless enough is coming in to supply the wants of the colony.

TIME FOR EXTRACTING.

I consider the proper time for extracting is when the bees commence capping over the cells. There is a diversity of opinion on this question, some apiarists going so far as to wait until the honey is all sealed over before extracting; while Mr. L. C. Root, whom I consider good authority on this point, recommends extracting while the honey is yet thin or raw, or in other words, extract every two or three days during the principal flow of honey. Mr. Root has described to me his method of evaporating and curing his honey, which is very simple and cheap. He claims several points in favor of his method, among which is the prevention of solid granulation. I think his method is worthy of a trial by those who manage their apiaries for extracted honey.

SEPARATORS.

Every bee-keeper who manages his bees for comb honey should use separators of some kind between the section boxes. Some apiarists argue against their use; but the bee-keeper, who strives to place his honey on the market in such shape as to command the very highest price, will find their use an absolute necessity. At the present time they are made mostly from tin. Many, however, have discarded these, and in their place have adopted those made from wood. From my own experience, I *much prefer* wood separators, *perforated*, to those made of tin, but as apiculture is rapidly progressing, we know not what will present itself in the future. In the way of separators, something new is about to be used: I refer to "Jones's perforated metal." It combines all the good qualities of the tin separators, the principal qualities claimed for those made from wood, and other good points not found in either.

In fact, I believe this "perforated metal" is destined to become one of the most useful articles used in the apiary; being already used in selecting the drones in fertilizing our queens, in the form of entrance guards, which is also a partial preventive against robbing, and will prevent swarms from absconding, thus abolishing the wrongfully practised method of clipping the queens' wings. By its use, I believe we shall be able to secure surplus comb honey in the centre of the brood-chamber, etc.

INCREASE OF COLONIES.

The amount of surplus honey obtained (whether comb or extracted) depends upon how the apiary is managed in regard to increase of colonies. As I am discussing the most practical method for securing the largest amount of surplus honey, I will not say anything about the *best way* for increasing our colonies. Those wishing bees instead of honey will change their manipulations, in conformity with the object in view, which should be different from the commencement of the season. The bee-keeper should at all times know the condition of each and every colony, especially at this season of the year. Much depends upon such a knowledge in order to obtain the best results. Most of our contributors to the various journals advise making an examination of each colony once a week. This is well enough for the inexperienced; but the bee-keeper who can justly claim the title of *apiarist* should at all times know the condition and progressiveness of each colony, from the outside appearances; and I have grave doubts if they who *cannot* do this will ever make a success of apiculture.

If the object is extracted honey, I would advise but little, *if any*, increase in colonies. With a proper use of the extractor, swarming may be controlled. With comb honey it is different. The number of new colonies should be in accordance with the season. By preventing any increase during a short flow of honey, or during a season below an average one, quite a satisfactory amount of surplus can be secured; while with a good season, or a prolonged flow of nectar, the best results can be obtained with a moderate increase of colonies. As I have said before, we should endeavor to have our colonies, as near as possible, of equal strength by the first of June; or at the commencement of the bloom from which our surplus is gathered. So we must labor to have our new colonies in the best possible condition and in every way equal with the old colonies; all being done, without diminishing the numerical strength, or in any way deteriorating the working capacity of any colony. At this writing, I consider the following method the most practical, and the one giving the best results.

Take an empty, or new hive, filled with frames of comb foundation, and a broad-frame with section boxes on each side of the brood-chamber, and place it in a position where you wish the hive to remain. Now take a frame of foundation, go to a colony and exchange it for a frame of brood and the adhering bees; put this frame of brood and bees in the new hive. Now go to another colony, and exchange a frame of foundation, for a frame of brood and bees, as before;

and proceed thus, until you have the new hive full of frames containing brood and the adhering bees. By the time the last comb is thus inserted in the new hive, the colony thus formed will be in a condition to receive any queen you may wish to introduce, which is done by simply letting the queen run down between two combs. Now carefully put in position the top surplus arrangement, and all is done; and you have a colony that is in a perfect condition to take every advantage of the bloom already secreting a bounteous supply of the coveted nectar.

In going through with the above manipulations, great care must be taken that the queen in the old colony is not taken away with the frame of brood and bees; also, if the colony can spare more than one comb of brood and bees without detriment, the apiarist should take two or more, as his judgment dictates.

If, however, by some oversight of the apiarist, a swarm should issue naturally, it should be returned, and then take every alternate frame of brood and bees from them (putting in their place frames of foundation) and place them in a new hive, alternately, with frames of brood, etc., taken from other hives. We should continue exchanging frames (of foundation, for those containing brood, etc.), as above stated, as often as our colonies become crowded for want of room to work; which, if neglected, will in most cases, induce the swarming fever, and thus greatly retard work in the surplus boxes. These conditions can be detected from outside indications, and the progress made in the surplus chamber, without disturbing the colony.

The skill and judgment of the bee-keeper should be taxed to his extreme mental and physical capacity in such observations, and should he be a quick and close observer, it will prove of great value to him in more ways than one, besides saving a vast amount of labor. The reader will readily see that, by adopting this method of securing an increase, every colony will be in a prime working condition, and if practised at the proper time, no swarming fever will occur; besides all this, our combs are built cheaply, quickly and without wasting any time in filling the hive with comb, before commencing work in the surplus boxes.

There are various other ways of managing this increase that are practised by some of our most scientific apiarists; but the one described is that which I believe to be *the best.*

I will, however, mention a method which I think is worthy a trial at least. Mr. D. A. Jones, who claims it a success, has put it into practice and claims, that by dividing the brood-chamber, in such a manner that the workers may have free access to both parts, while the queen will be slower in finding her way between the two compartments, swarming may be controlled. This is done as follows: insert one or two broad-frames, containing surplus boxes filled with foundation, in the centre of the brood-nest, using as separators the " perforated metal," described elsewhere, and cut the full size of the broad-frame. It is best to use the metal as division boards, placing the broad-frames at the sides, in the centre of the hive, after spreading the brood. Thus the bees are allowed to pass from one side to the other, at any point, while the queen must pass either over or under these broad-frames, to enter the one part from the other.

At the end of the season for white honey, we find that we have a number of boxes that are unfinished. If they are one-half or more completed, they can be easily finished. Many, however, prefer to *extract* the honey from those that may be unfinished at the close of the season, allowing the bees to clean up the combs, when they are stored away for another year. I think this the better plan, as these boxes filled with comb offer excellent inducements for getting the bees in the boxes the next season. If the bee-keeper concludes to have the best of them finished up, he must do so by *feeding back*. To make this profitable, several colonies should be selected for the purpose. They should be strong in numbers, and the combs containing brood should be exchanged for others that are all, or nearly all, solid honey. Those using broad-frames for side boxing should remove them, and in their place arrange the feeders, having the unfinished boxes on top. As a feeder for *such purposes*, I know of none better than "King's Automatic" which is made to hang in the hive, the same as broad-frames. If our hives are not arranged for side storing boxes,

8

we can remove the frames containing brood, etc., and crowd together the frames containing the honey, thus leaving room at the sides to suspend the feeders. Mr. Jones, who uses a stationary bottom-board to his hives, feeds by simply tipping the hive back a little, and pouring the food on the bottom of the hive, from the entrance, by means of a funnel. This is the best feeder for those using stationary bottom-boards.

CARE OF HONEY.

Every apiarist should have a small building or a room in a larger building exclusively for storing and for the proper curing of his honey. This room should be well ventilated by means of two small windows, or one large one protected on the outside by fine wire-cloth and absolutely moth- and mouse-proof, and he should use care, while passing through the door, not to allow any millers to enter the room. If spiders and large ants should take up their abode in the room, have no fears as they will do no harm but will destroy all injurious insects that make their appearance, and take up what little leakage there may be from the honey. On the floor lay down strips $\frac{1}{4}$ of an inch thick by $1\frac{1}{2}$ inches wide, and far enough apart to allow the boxes to rest upon these reaching from the centre of one to the centre of the next strip. Set a single tier of boxes on these, then lay over them pieces of manilla paper to catch the leakage from above, and on top of this, place more strips and another tier of boxes and so on until you have seven or eight tiers of the two-pound boxes.

As fast as the boxes are taken from the hives they should be put in trays or crates for the purpose and carried into the honey-room at once. Do not leave the honey exposed to the moth-miller at any time. Herein lies the secret of evading the ravages of the moth.

Extracted honey does not require so much care and attention as comb honey. The principal point is to have the honey properly cured or ripened, when it may be put in barrels or casks and stored in a room with a moderately high and even temperature. All cellars should be avoided for storing honey of any kind. If the honey is not extracted before it is sealed or capped over or nearly so, it may be put in the receptacles immediately; but if the honey is extracted while it is yet thin and raw, it should be cured or evaporated before being put into packages for shipment. I think this is best done by Mr. L. C. Root's method, which is illustrated and described in the Proceedings of the Northeastern Bee-keepers' convention and published in the "Bee-keepers' Exchange" for March and April, 1883.

SULPHURING.

We have been large producers of comb honey for the past twenty years, and as yet have never sulphured our honey and have never been troubled with moths. If the above precautions are heeded, there will be no need of resorting to brimstone to protect our honey against the ravages of the moth. As the market for honey seldom opens before the latter part of September or first of October, the honey should remain as we pile it in the room, until just before this time.

PREPARING FOR MARKET.

The first work in preparing our honey for the market is to clean all propolis, wax, etc., from the sections. Then comes

GLASSING.

The glass should be nicely cleaned and put on the sections with tin points or white glue. If with the former they should be set one side from the centre of the box and in an opposite direction from the one on the other side. Great care must be taken not to crack or start the combs by using glass too large as it will cause the honey to leak and will weaken the combs making it more hazardous in transportation to market.

I am endeavoring to improve on the method of fastening the glass on the sections by using strong labels to cover the top of the section and reaching one-fourth or one-half inch down on each glass, the bottom to be fixed the same way.

In conversation with Mr. D. A. Jones, he thought he could furnish labels similar to those he now makes and uses on small pails for extracted honey that would give entire satisfaction. If so, it will add greatly to the neatness and attractiveness of our packages, covering the stains from propolis and wax, and at the same time contain an advertisement for the producer on every box of honey.

The one-pound sections should not be glassed but nicely cleaned and put into those neat paper boxes as made and used by Mr. I. L. Scofield. These boxes contain the producer's card and are weighed as so much honey, consequently the cost is nothing. With the present method of crating, the producer is out of pocket on the cost of glassing the crates, which is not weighed as honey and is therefore a dead loss.

CRATES.

Our crates, or cases, should be made with all the accuracy we exhibit in making our sections. White basswood and poplar have several points in their favor, over other woods, as the best material. Care should be taken not to stain or soil them in handling. They should hold twelve of the two-pound sections or twenty-five of the one-pound boxes. After the crate is made the bottom should be fixed in the following manner: take strong manilla paper and cut one inch larger each way than the inside of the crate. Fold this over a board (made for a pattern) that will just fit inside the crate, turning up the corners, thus forming a paper pan that will just fit in the bottom of the case. After the pan is in position tack strips of wood three-sixteenths of an inch thick and seven-eighths of an inch wide on top of the paper and crosswise of the crate, placing them far enough apart so that the boxes will reach from the centre of one to the centre of the next strip thus forming a space between the bottom of the section and the paper pan that will catch and *hold* all the leakage. In this way the boxes are free from any daubing from leakage of the honey. I consider this a great improvement; although the inventor, Mr. Samuel Snow of this place, has used the same for several years, it is known to but few as yet. It should be known to every producer of comb honey.

GRADING AND CRATING.

The crates should first be weighed and marked in small figures with a pencil. The most convenient way to crate is to have as many crates by your side as you have grades of honey.

There should be two grades of white honey and two grades of dark. The best grade should be marked XXXX and composed of sections with all straight and even combs all sealed over and the cappings or comb stained but little if any. Mark the second grade XXX, which should comprise all boxes of white honey not fit to be put in the best grade. The the third or XX grade is made up of sections that are one-half or more white but finished up with dark honey. The fourth or X grade takes all the dark and buckwheat honey.

In putting the sections in the crates it is rulable to put the best side of the outside boxes out, but the boxes inside should average as good as those on the outside. A great deal of the honey on our markets is crated so that nice white honey appears on the outside while the inside is made up of mixed honey. This is called *veneering* by the dealers and is the cause of many complaints and denunciations on the part of the bee-keeper, who thinks the dealer has wronged him by selling too cheap when really no one is to blame but the apiarist himself. Honey put up in this way must be sold as mixed honey. The bee-keeper who is so devoid of principle as to practise this should receive no sympathy from any quarter.

I wish here to speak of another bad custom and one which I have entirely overcome in this locality the past season. In buying honey, the buyer makes a difference of from two to four cents between each grade and makes his contracts in that way; this makes a difference of about five cents (on a two-pound section) between the two grades, and when crating many are inclined to look at the five cents, more than at the strictness of the grading. By selling both grades of white honey at the same price and the two grades of dark honey at another price, the producer finds nothing to warp his judgment; consequently, all grades are far better and give better satisfaction to all concerned. In crating the one-pound sections the

crates should be made for a single tier only, taking five
boxes in length and five wide. The middle box on each
side should be glassed on one side, so as to show the
average quality of honey in the crate, as by using the
paper boxes referred to previously, the honey is not
visible.

WEIGHING AND MARKING.

It is a common practice with bee-keepers to put
their name and address, as well as the gross weight
and the tare, on each crate, with a stencil plate. This
should not be done. Put your name and address on
each section if you like but not on the case. Put
nothing on the crate but a mark denoting the quality
or grade of the honey and the *net* weight neatly marked
on one end in stencil figures. The crates should
weigh even pounds *net; i. e.*, there should be no
fractions of pounds.

MARKETING.

Many bee-keepers succeed in obtaining a large
amount of surplus, but comparatively few seem to
realize "top prices" for their honey, either from a
lack of knowledge of the requirements of the market,
or their inability in supplying the demands of the
consumer.

Some apiarists take extra pains to have their crates
and sections neat and clean, and all that could be
desired in attracting the attention of the consumer,

but spoil *all* by an improper system of grading; others bend their whole energies in the direction of perfect grading and beautifying the outside appearance of the crates but lose sight of that most important point of having the sections clean, bright and free from all propolis, etc.

We must cater to the wants of the consumer, and not so much to some of the fancied caprices of the dealers or middlemen, and while we endeavor to please the public we should not disregard our own interests financially and otherwise. Dealers are constantly on the alert to increase their profits; and by managing to induce their bee-keeping friends to put up their honey in some new style of box or crate, or to fix it up in some way so as to make it a novelty, they succeed in increasing their profits at the expense of the apiarist and at an extra cost to the consumer, with a tendency of lessening the consumption of our products. Never deviate from the two sizes of boxes I have mentioned. Strive to place them on the market in the very best shape possible, and your honey will please the consumer in all that the most fastidious connoisseur could desire, and command the very highest market prices.

Always supply your home market before looking elsewhere. If you have a surplus after that, ship it where it will net you the most. Always sell for cash when you can; it is better to sell for two cents per pound less than what is quoted by commission dealers, than send it on commission, as it costs the consignor that difference in paying the various bills of expense, besides the risk of losing the entire consignment by the failure of the commission firm.

If, however, you are forced to send on commission, ship to a firm that makes a specialty of our products and is known to the bee-keeping fraternity. Nothing will frighten the retailer so quickly and lessen the prices more than the practice of consigning to any and all of the hundreds of commission houses. To succeed best we must concentrate our products on any market. Our bee magazines could help us much more than they do in this matter.

SHIPPING.

This has been one of the perplexities in marketing our products, yet it is quite simple and when well understood comb honey can be safely shipped to any part of the world. My long experience, crowned with the best success attainable, will verify the assertion that what I shall say on this subject may be considered as authority. First, secure a good clean car. By presenting your freight agent with a little honey you will have the best to be had; before loading, the car should be swept clean and paper spread on the floor to keep the coal cinders from soiling the crates. Commence loading by placing a row of crates against one end of the car with the combs running lengthwise with the track. The crates should not be piled more than three feet high. At one side of the car we will have a vacant space of a few inches; this should be fixed by putting a board against the sides

of the crates and stay-lath it so that the crates cannot
move ; now place another tier of crates against the one
just put in and fasten at side as before ; proceed in
this way until your honey is all loaded, or you have
filled the car as far as the doors. Here fit good strong
boards against the ends of the crates and stay-lath so
that there is no possible chance for a single case to
move. The other half of the car should be managed
the same way and when the car is thus loaded it will
not be far from ten tons net weight. Over the tops
of the crates tack old newspapers to keep the dust
and cinders from sifting among the cases. In this
way the honey may be sent to California and back
again without breaking a half dozen combs. Always
do your own loading and reserve the privilege of
superintending the unloading. Always ship by freight
when you can. The charges are much less and the
risk no greater than by express.

EXTRACTED HONEY.

If sold in our larger markets or shipped away from
home it should be put up in kegs holding about one
hundred and sixty pounds net. On one end mark with
stencil plate the producer's name and address, the quality of the honey and the net weight. Honey put up in
five and ten gallon tin cans does not sell so readily as
that put up in kegs. If near a large town, I should put
my extracted honey in one, two and a half, five and

ten-pound tin pails labelled with Jones' fancy labels
for that purpose. At present I cannot endorse putting
up extracted honey in packages of less than one pound.
From late experiments I am satisfied that cheap tin
should be avoided as the best chemists have shown
that honey in such tins is tainted with poisonous acids
or minerals.

I will here venture on rather delicate ground knowing that I shall meet with strong opposition by some,
but nevertheless I shall venture the assertion. The
finer or white grades of extracted honey find ready
sale at remunerative prices. But it is a stubborn fact
that the darker grades are a *drug* on our markets.
Time after time the question has arisen "How can we
establish a trade or how can we find an outlet that
will consume all of such grades of honey?" Grape
sugar is now largely used in the manufacture of beer.
In conversation with a prominent brewer of Brooklyn,
who is now using dark extracted honey in place of
"grape sugar," he said "that the honey could be used
at eleven cents per pound and be cheaper than 'grape
sugar' besides making a beer that could not be
excelled in purity and healthfulness." If bee-keepers
will go to a little trouble they can establish a trade in
this line that would demand more honey than is now
produced. I do not wish to be treated to a "temper-
ance lecture," so will suffice by asking, *Is it not far
better to give the consumer of beer a pure and good
article than an adulterated one?*

SIZE OF BOXES.

Much may be accomplished in the way of regulating the size of section boxes. We will first take up the two standard sizes found on our market to-day and show what the difference in price must be to make them equally profitable to the producer. It is admitted by the best apiarists who have had the longest and largest experience with the one-pound boxes that the amount of surplus honey secured is $\frac{1}{8}$ less than with the two pound sections. Therefore we shall take that as a standard. In producing and preparing for market 1000 lbs. in one-pound sections it will cost, for 1000 boxes, $4.50. To fill these with foundation, it will take 10 lbs., 10 ft. to the pound, which at 60 cts. is $6.00; 40 crates for crating the same at 15 cts., is $6.00 more; 80 glass for glassing sides of cases at 5 cts. makes $4.00; thus for the 1000 lbs. it costs $20.50.

1000 lbs. in sections $5\frac{1}{4} \times 5\frac{1}{4} \times 2$ in. will cost for 500 boxes $3.00; to fill them with foundation 10 ft. to the pound will require 7 lbs. at 60 cts., which is $4.20; 40 crates for crating same at 9 cts. makes $3.60; $10.80 in all and a difference in favor of the two-pound section of $9.70.

To glass the 500 two-pound sections, it will take $3\frac{1}{2}$ boxes glass at $2.40 which would be $8.40; this glass weighs 60 lbs. to the box or 210 lbs. in all, which is weighed as honey, and at 20 cts. would be $42.00; it costs us as above $8.40, leaving a balance in favor of the two-pound sections of $33.60. By having the honey stored in the two-pound box, we gain $\frac{1}{8}$ or 150 lbs. honey, which at 20 cts. is just $30.00 and making a grand total in favor of the two-pound sections of $73.30 or a difference between the two sizes of $7\frac{1}{3}$ cts. per lb.; therefore, to be as profitable to the producer, the one-pound box must sell at $27\frac{1}{2}$ cts. when the two-pound section is selling at 20 cts. I have said nothing about the labor in glassing the two-pound boxes, etc. The extra labor required in making double the amount of boxes, and that performed in the various manipulations of the one-pound sections, would leave a nice margin on labor in favor of the larger section.

I will now consider the one-half pound boxes, a subject which is at present agitating a "certain class" of bee-keepers and compare *them* with the two-pound sections.

At our late Northeastern Bee-keepers' convention, it was acknowledged by those apiarists who have had more experience than any others with the one-half pound boxes, that they could get but $\frac{2}{3}$ as much honey stored in them, as they could in the one-pound box, and then they could succeed only when honey was the most plentiful. For 1000 lbs. honey in one-half pound boxes it will take 2000 sections at $4.00, which will be $8.00; to fill them with foundation it will require 20 lbs. 10 ft. to pound at 60 cts., making $12.00; 50 crates for crating same at 15 cts., will make $7.50; the 100 glass for sides of cases at 5 cts., will be $5.00 more and a total of $32.50. If we can get but

⅔ as much in the one-half pound as we do in the one-pound box; and but ⅗ as much in the one-pound as the two-pound sections then while we secure 1000 lbs. in the one-half pound box we can obtain 1715 lbs. in the two-pound sections. The cost of this will be, 850 boxes at $6.00, $5.10; 12 lbs. foundation 10 ft. to pound will fill them and at 60 cts. will be $7.20; 70 crates to crate the same at 9 cts. equals $6.30, making in all $18.60 or a difference in favor of the two-pound sections of $13.90. To glass the 850 boxes it will require 6 boxes glass, at $2.40, making $14.40; weighing 60 lbs. to the box, makes 360 lbs., glass at 20 cts. (being weighed as honey), we have $72.00, or a balance on glass of $57.60 in favor of the two-pound sections. The difference in amount, produced as before stated, is 715 lbs. which, at 20 cts., will make $143.00 more and a total of $214.50 or 21½ cts. difference in the two sizes.

Therefore the one-half pound section must sell for 41½ cts. when the two-pound boxes sell at 20 cts. to be equally profitable (financially speaking) to the apiarist, to say nothing about the vast amount of extra labor and time lost in manipulating the one-half pound sections which would put the price at about 50 cts. to be a success. The quotations then should read like this: two-pound sections 20 cts.; one-pound boxes 28 cts., one-half pound sections 50 cts. per pound; then the one-half pound box would bring 25 cts., the one-pound box 28 cts. — a difference of only 3 cts. — and the two-pound section 40 cts.

It is not the consumer that demands the one-half

pound package or the one-pound box. It is the dealer
who is constantly endeavoring to introduce something
new in the line of a novelty, at the expense of the producer and consumer alike. They are assisted by our
supply manufacturers, and they too have an axe to
grind as it increases their sales and consequently their
profits.

It is time that apiarists viewed this matter in its
true light and I trust the reader will pause and
consider before his business is ruined. I will venture
to predict that the time is close at hand when the one-
pound section will sell at the same price as the two-
pound box. If the one-half pound box is produced in
large quantities, it will be but a few years before they
too will find their level with the two-pound section.
The $5\frac{1}{4} \times 5\frac{1}{4} \times 2$ section will be the standard in the
near future. *Wait and see.*

COÖPERATION.

Nearly every apiarist in the country must know
that I have worked earnestly for the past several
years to demonstrate to the bee-keepers of America
the many advantages to be had and the large amount
of money that can be saved by united action. In
this I have been opposed only by those interested in
the manufacture or sale of supplies, and their friends.
Of late I can see the tidal wave steadily moving on-
ward that is sure to accomplish the desired end.

Mr. T. G. Newman in his work on "Bees and

Honey," published in 1882, tells us that "any method, that will add one cent per pound to the marketable value of our products, is worth to the producers three-quarters of a million of dollars; and any error of management, causing a reduction of one cent per pound is to them a corresponding loss."

In addition to what Mr. Newman has said, I will add, "any method that will *save* one cent per pound in preparing our products for market is worth to the bee-keepers of America nearly three-quarters of a million of dollars; and a *failure* to save that one cent per pound is to them a *loss* of just that amount."

This amount can be *saved* by buying our glass, sections, hives, foundation, literature, and the many implements of the apiary at manufacturers' prices. A *like* amount can be *added* (by a majority of bee-keepers) by united action in selling our products, but to accomplish all this requires organization. Reader, if you will express your willingness and lend assistance in this matter, we shall soon advance one step nearer perfection.

There are other points of material interest to the apiarist that can only be accomplished by coöperation, such as collecting accurate statistics; providing a relief fund for the benefit of our families after our death; perfecting a system in grading and marketing our products and bringing to bear, on our railroad officers, legislatures and government officials, an influence that will command due respect and undelayed action when our interests are involved.

9

CONCLUSION.

The reader who has perused these pages with care will notice that I have not described the various conditions and the many things that would greatly facilitate the various manipulations of the hive, nor even mentioned the many methods pursued by different apiarists in obtaining their surplus and their management during the swarming season; indeed, it would take a large volume to do this. I might, perhaps, have gone into more minute details for the benefit of the novice, but as this work is not an A B C book, but a volume whose pages are filled with the results of years of experience, I have endeavored to present to the reader the latest and most scientific and practical method of management to secure the best results from the apiary. How well I have succeeded, time, that great leveller of all things, will tell.

Fayetteville, N. Y., March, 1883.

THE NEW RACES.

BY SILAS M. LOCKE.

APICULTURE, as a science or vocation, received almost its first impetus from the combined efforts of the lamented father of practical apiculture in America, Moses Quinby, and L. L. Langstroth (the inventor of the movable-comb hive) ; both, authors of works, which constitute the foundation of practical apiculture. Until this time bees had only been kept in the old-fashioned way, and with few exceptions in box-hives, straw skeps, etc., and the idea that apiculture would ever assume the dignity of a science, or become established and remunerative as a vocation, was not conceived. True, marked and important advances had been made in Europe, but it was left to America to establish the first practical methods of bee-keeping.

THE ITALIANS FIRST IMPORTED.

In the year 1853, Dzierzon (aided by the Austrian agricultural society at Vienna) succeeded in importing his first colony of the Italian bees, and after thoroughly testing them pronounced them superior in every respect to the native bees, as they swarmed earlier, were more industrious and hence better honey gatherers, more gentle and yet more courageous and active in self-defence and far more beautiful.

(131)

ITALIAN BEES FIRST IMPORTED INTO AMERICA.

When the news of this fact was wafted across the water and these wonderful and encouraging reports were corroborated by other European apiarists, many prominent and influential bee-masters on this side of the water became deeply interested in them and several attempts were made to import them, all of which, however, proved unsuccessful until the year 1859, when Messrs. Wagner, Colvin, and Mahan succeeded in making the first living importation. Although the pioneers of apiculture had laid the foundation, the advent of the Italian bee marks prominently the starting point of the "bee-fever" (so called), and an interest began which has continued until apiculture has assumed an importance which places it side by side with agriculture, horticulture and kindred industries, both as a science and remunerative vocation, an interest which should never cease until our national government recognizing the importance of such knowledge shall appropriate means for its dissemination, and the educational bureau and colleges shall admit the necessity of teaching it as an essential branch of education. This may and should be recognized here, as in other prominent nations.

For the last twenty-three years the Italian bees have been thoroughly tested and experimented with in this country, and although it is now generally admitted that they are not a distinct race, yet the opinions and statements of European bee-masters have been verified and the Italian bees have gained much credit and have proven in many respects a superior race, and careful selection and breeding have produced the American Italians, which are far superior in every respect to any ever imported from Italy. Their markings and characteristics being well known it is not necessary to allude to them here.

THE FIRST NEWS OF THE CYPRIAN BEES.

A number of years after the first importation of the Italians, the news came from Europe that another and superior race of bees had been discovered on Cyprus Island, in the Mediterranean, and soon after the news of their importation into Europe by Chancellor Cori of Bruex, Bohemia, and Earl Rudolph Kolowrat-Krakovsky, frequent and interesting articles regarding their excellent qualities were translated and published in our leading journals from the above and from the writings of Gravenhurst, Stahala and others, men of worth and ability, whose opinions and statements command respect; and in almost every instance the universal opinion concurred with that of Gravenhurst, who said: "Count Kolowrat-Krakovsky, Chancellor Cori, Hilbert, Stahala and I, have written on our banners, 'Cyprian bees everywhere.'"

THE CYPRIANS FIRST IMPORTED FROM EUROPE.

Previous to this, much had been said about the "*coming bee*," and the bee-keepers became electrified with the hope that in this new race or its crosses with the Italian was centred the "Apis Americana," and true to American enterprise and energy they never rested until several attempts were made to introduce them. Mr. Julius Hoffman of Fort Plain, N. Y., succeeded in obtaining some of the European stock, and it is stated that Mr. Blood of Quincy, Mass., also received an importation; *but this was not enough.* In order that we might be certain of their purity they must be imported direct from Cyprus Island. The interest continued to increase, and about this time news of the Holy Land bees came to us; at last Mr. D. A. Jones, the bee king of Canada, a man who had the interest and advancement of apiculture at heart, secured the services of the well-known and enthusiastic bee-master, Frank Benton, and together they set sail from New York city (Jan. 1, 1880) on a voyage of more than six thousand miles in search of the new races. I will leave them here, passing over the history of their travels as it has been published in the journals.

THE IMPORTATION OF THE CYPRIAN AND HOLY LAND BEES BY D. A. JONES.

Early in June, 1880, Mr. Jones reached his home in Beeton, Ont., with one hundred colonies of Holy Land and Cyprian bees, and that year will mark one of the most important and prominent eras in the history of apiculture. Much credit is due Mr. Jones for presenting to Europe and America the first living specimens of the Holy Land bees, and also for first bringing to *our* country Cyprian bees direct from Cyprus Island, procured at an enormous expense as well as great sacrifice of personal comfort; and to Mr. Benton for the enthusiastic and untiring energy with which he assisted in prosecuting the work which has given to America two of the most important races of bees extant; one of which (the Holy Lands) is, as far as can be known, the first race given to man.

Since the importation of the Cyprian and Holy Land bees, they have been introduced into almost every state in the Union and over a large portion of Canada, and they have been thoroughly tested; giving results concurring almost invariably with the statements of European bee-masters.

THE MARKINGS AND CHARACTERISTICS OF THE CYPRIAN BEES.

Cyprian bees are in many respects similar to the Italian, and yet they have many distinctive features by which they plainly differ from them, viz. : all pure Cyprians are more or less yellow on the sides and under parts of the abdomen, the tip or point being very black, the yellow part underneath their bodies looking as glossy as though varnished; in fact, when the hive is opened and the rays of the sun penetrate between the combs, the golden hues of the bees reflected back present a sight that is simply beautiful, and this is one important test of their purity. In form they are somewhat smaller than the Italians, quite slender and wasp-like, with sharp pointed bodies.

I have known the Cyprian queens to pass through an aperture, through which those of the Italian, Holy Land or black races, could not pass. The hair covering of the Cyprian bee is a lighter yellow and of a lighter shade than that of the Italians. The first three rings of the abdomen are yellow (or orange color) the entire width, while among the Italians many of the workers have only two narrow ones and sometimes only one. There are, however, Italian queens which produce all three-banded workers which are uniform in their markings.

The posterior rings of the bodies of Cyprian workers are broader than those of the Italian, and when examined you will notice that the upper portion is partially black, and in terminating on the sides a perfect halfmoon (there are generally two of these) is formed and also that there is no intermingling of color. With pure Cyprian bees this is an invariable and uniform marking.

The Cyprian bees are more yellow and more beautiful than the Italians, and we have every reason to believe that by careful selection and breeding they will become still more marked in this respect. The Cyprian queens are generally longer lived than the Italians; most of the latter outlive their usefulness regarding prolificness in one or two years. Again, the Cyprian queens (as well as the Holy Lands) are wonderfully prolific, laying almost an incredible number of eggs in a season, notwithstanding they are smaller and more slender in form than the Italians. The Cyprians build less drone-comb than either the blacks or Italians. When necessary, they fly farther in search of stores. They are strong, quick-motioned and very active, and fly very swiftly, more so in fact than the Italians. Their bodies are telescopic, which enables them to carry large loads of honey, and to contract at other times to quite small proportions. The Cyprian bees are more hardy, and hence winter better than the Italians, come out better in the spring and do not dwindle so badly. They are very energetic and determined in self-defence and in protecting their stores against robbers, so much so that they seldom allow robbers to alight, but often meet and seize them on the wing; in fact, I have found them generally proof against robbing. They are more judicious about leaving the hives in unfavorable weather and have a keen and more acute scent than either blacks or Italians. They do not stop breeding as early in the fall after the first frosts as do the blacks and Italians, but keep on breeding, thus better fitting them to stand the long, cold winter and spring months, so far as having young bees is concerned.

Pure Cyprians invariably have a beautiful golden shield between the wings on the thorax. The Cyprian drones are very handsome, more so in fact than those of any other race. Cyprian bees can be transported more easily than Italians or blacks as they do not gorge themselves so freuqently or readily and can easily contract their bodies.

The Cyprians have gained the reputation of having a choleric disposition, being easily irritated, and revengeful to the last degree. Regarding this matter opinion seems divided, as some of the most noted

European writers state "that with proper management
the Cyprian bees can be handled as safely as the
Italians." Messrs. Jones, Benton, and many others,
both in Canada and the United States, agree with
them in this opinion, with the exception that there is
occasionally a colony which defies manipulation, under
the mildest treatment. The reason why this is so is
quite evident. They have been confined upon an island
where forage (generally speaking) is so scarce that
"Mr. Benton wonders that there is a bee left in such
a country ; and claims that where the Italians or blacks
could not survive the Cyprians will live and thrive,"
and further claims that only the fittest have survived.
The untiring energy and perseverance necessary to
maintain a subsistence under such circumstances have
developed their characteristic activity and restlessness
as well as fitted them for better defending their stores.

It has been said that after being kept for a few
years in Italy, Austria and Germany, where forage
was plentiful, and subjected to careful breeding and
handling, they gradually lost much of their irritabil-
ity, and hence became better natured. May this not
hold good in our own country? Since their introduction many poor hybrid Cyprian queens have been
sent out all over the country as *pure* Cyprians,[1] and in

---

[1] It will be remembered that queens reared from the first imported Italian stock were not uniform in color or markings, many being totally black,
while others were beautifully yellow or golden. One queen would hatch and
prove beautiful in color and uniformly marked, while her sister from an
adjoining cell would be wholly black. I have experienced this same difficulty with the Cyprians, and although the queens of this race are not as handsome or beautiful in color as the golden Italians, the former being generally
striped and often quite dark, yet after a few years of careful selection and
breeding they will be purely bred by all, more uniformly marked and better
natured.   [ED.

fact, even at the present time, very few bee-keepers can compare, point out, and describe the essential markings of the queens, workers, and drones, of the Cyprian and Holy Land races.

The dispositions of the Cyprian bees may be faulty enough, but it is to be hoped that this may be overcome by careful selection and breeding, and the change in climatic influence, as they far surpass the Italians as honey gatherers, breeders, and in cell-building.

Both the Holy Land and Cyprian bees are noted for the large number of queen-cells that they will build and care for. The opinion has been advanced (and I think that it has reason on its side) that the Cyprian bees, like the Italians, are not a distinct race, but descendants of the Holy Lands, and have been confined to this island until they have developed characteristics which give them the appearance of a distinct race.

Be this as it may, when we have given the Cyprians a fair trial under careful management, we shall find them an important addition to the races which we now have.

THE HOLY LAND BEES.

The Holy Lands (or, as the natives call them, the Holy bees) are found in Mount Lebanon, Mount of Olives, Mount Hermon, the valley of Sharon, Bethlehem, the hills of Judea, Jerusalem, Jordan, Ammon, East of Jordan near the desert, Galilee, Damascus and various other places in Syria. They have existed thousands of years, evidently from the foundation of the world, and have had no chance to mix with other races; the monks in the old convent near Jerusalem believe that they are the first bees given to man.

There is but one distinct race throughout the different portions of the Holy Land, although they vary somewhat in minor characteristics, and we have every reason to believe that we have at last reached the starting point, or fountain-head of superiority; that they are a distinct race and that many of the yellow (or golden) races may be traced to the Holy Land bee for their origin. They are in very many respects similar to the Cyprians, while they have some prominent characteristics and markings, by which they may be distinguished from any of the other races.

THE MARKINGS AND CHARACTERISTICS OF THE HOLY LAND BEES.

Like the Cyprians, they have the golden shield between the wings on the thorax and the black half moon previously described (the latter marking may be seen better when the bees are filled with honey and preserved in alcohol); they are strong, energetic and active, and although not as irritable or vindictive as the Cyprians, yet they resent rough or careless handling much sooner than do the Italians. They are swift on the wing, flying so rapidly that only those heavily laden with honey can be secured on the bloom, and fly farther in search of forage, when necessary, than the other races; again the Holy Lands have been known to work in large numbers on red clover (large heads) when there was a plentiful flow of white clover honey.

The hair covering of the Holy Land workers is lighter than that of the Cyprians or Italians and they have a thicker coat of it. The alternate rings of yellow and black are so divided by the light hair covering, that the bees appear to be beautifully striped, presenting an entirely different appearance from any of the other races when clustered on the combs, running around on the front of the hives, or on the wing.

When the hive is opened and the combs examined, the young (just-hatched) bees huddle together in small clusters and tumble from the combs if jarred; notwithstanding some of the young bees are strikingly small, yet in a few days they become fully as large as the rest; characteristics not found in either of the other races. The Holy Land queens are smaller than the Italians but larger than the Cyprian, not being quite so slender, and are fully as prolific as the latter. The Holy Land bees are indisputably the best race for cell-building, and if permitted will build an unusually large number of cells and take good care of them.

The Holy Land queens are generally quite yellow while the Cyprians are very dark. The Holy Land bees (like the Cyprians) winter better than the Italians, come out better in the spring and do not dwindle so badly; they are fully as energetic in defending their stores against robbers and gathering honey as the

Cyprians, and both the Cyprians and Holy Lands will make a greater effort to obtain honey when there is a scarcity, than the other races.

Everything considered, we regard the Holy Land bees the most superior race ever imported into our country and that when they have been subjected to the same careful selection and breeding as have the Italians, they will command a far more prominent position among our races than have the Italians.

THE APIS DORSATA.

Very little is known of the true merits and value of this new race of bees ("the *Apis dorsata*" or bee of Ceylon). True, a number of very interesting articles have been written regarding them, some of which extol their good qualities and beautifully picture them sipping from the flowers large quantities of the precious nectar, inaccessible as yet to other races; but the facts in the case do not seem to warrant such imaginations even.

From all the information which we are able to glean regarding them; they seem to be rather indolent and shiftless, with no particular "care for the morrow," losing what little energy they do possess when made queenless, and carrying a weapon of defence, which would almost defy a coat of mail.

In 1881, Mr. Frank Benton, agent for D. A. Jones, made a voyage to Java and Ceylon, in search of this race of bees, and after many long and tiresome journeys through almost impenetrable jungles, alone, and thousands of miles from his home and fellow country-men, he secured two colonies of these bees; but, owing to exposure, over-exertion and the poisonous atmosphere, he contracted an almost fatal attack of the jungle malarial fever, and was unable to give proper attention to the preparation of the bees for their long journey, the result being that they died when they had nearly reached their destination. It is to be hoped that at some future day, other and successful attempts will be made to secure and fully test them. Much credit is certainly due Messrs. Jones and Benton, although unsuccessful, for the untiring energy and perseverance displayed in doing their part so well.

CHARACTERISTICS AND MARKINGS OF THE APIS DORSATA.

Mr. Benton, in a letter to Mr. Jones, calls them "wonderful bees as large as queens, blue-backs, with shining blue wings and orange-colored bands under them, having the appearance of great wasp-colored hornets, beautiful but dangerous looking, irritable and very tenacious when excited, and after becoming queenless they would take no care of the brood and soon dwindle away. Again, while very ferocious in their forest home, where nothing but smoke will subdue them, yet they can be handled with no fear and without smoke, when in movable frame hives, provided they are not jarred or breathed upon, and no quick motion made. They build combs four or five feet long and three or three and one-quarter broad, with about one and one-half bushels of bees to the swarm. They do not repair combs readily after being transferred, and seem to inherit many of the characteristics of our well-known *bumble bee.*

Mr. Benton also secured specimens of other races, but did not feel warranted in importing them as he did not consider them very valuable.

HUNGARIAN BEES.

Four years ago, Mr. Alley imported this race of bees (the Hungarians), and after testing them carefully, found them very fair honey-gatherers, quite gentle, as much so as pure Italians ; and the queens vigorous and prolific.

The queens, workers, and drones of this race are quite dark, although the hair covering of the workers from imported queens was whitish or light gray. When trying to keep them pure, he found that they degenerated and soon became nearly similar to the common black bee in color, but when crossed with the Italians, the workers were quite handsomely marked.

They proved unprofitable on account of a great propensity to swarm, and bees inheriting this quality or characteristic are of but little account to the apiarist, whose object is to secure the largest amount of surplus honey. Where increase of stock is desired, this race cannot be excelled.

THE COMING BEE.

There is no subject pertaining to apiculture in which I am more deeply interested, or consider of more importance to the bee-keeping fraternity, than the improvement of our races.

Many of our most intelligent, thoughtful and practical bee-masters are studying and experimenting carefully with the different races, in order to develop a strain of bees which shall include all the most desirable qualities and justly merit the title of the *Apis Americana*.

I believe, with them, that this can only be accomplished by careful and judicious selection and still more careful breeding. Further, I believe that no race or strain of bees extant possesses all the necessary qualities or characteristics.

As a people, we are often too sanguine and are not content to follow the straight and sure but slower path which leads to success. We too often jump at conclusions and then attempt to establish or prove them afterwards, instead of carefully investigating and thoroughly settling every point conclusively as we proceed; and never has any race of bees been long imported into this country before its merits and value were decided upon, — decisions that in many instances were withdrawn after thorough trial and testing of said race. For instance, many of the first Italians imported were faulty, and soon after their introduction into this country, many prominent apiarists, who to-day are the strongest supporters of their

10

good qualities, then pronounced them an inferior race
of bees, and no better than our German bees.

Again, it has been only about three years since the
Cyprians and Holy Lands were first imported; and
yet to-day many of those seemingly good judges,
pronounce them no better and perhaps inferior to
some of the races and strains which we now have,
while I am confident that time and experience will
demonstrate their mistake; and, further, that *one* of
these races, at least, is to figure largely in the
"coming bee."

THE NECESSARY QUALITIES OF THE COMING BEE.

The question now is, What shall constitute the race
or strain which shall sit enthroned upon this the
highest position among the races, and upon what
foundation shall we build? I would reply:

1. They must be a hardy race, able to withstand
successfully the trying changes and severe winters
of our northern climate.

2. Good breeders: keeping the hives well supplied
with brood and young bees from spring to fall and
even during the most trying portions of the season, as success depends largely on strong populous
colonies.

3. Gentle and quiet in their movements, thus
permitting of easy manipulation and this without
diminishing their working qualities.

4. Good honey-gatherers, energetic, determined,

and successful in their efforts to secure every drop of the precious nectar, which circumstances will permit, storing and capping the same in an attractive manner, and fully as energetic and determined in protecting their stores and homes against the invasion of robber bees.

5. Strong and active on the wing, and capable of making long journeys when necessary, without being exhausted.

6. Long-tongued, in order that they may sip the precious nectar, as yet inaccessible to the bees, from the many honey-producing flowers which now refuse to yield up their hidden sweets, wasting them on the summer air; and, finally, beautifully and uniformly marked, also duplicating the above markings, qualities and characteristics.

I would thus picture the coming bee, or *Apis Americana.*

THE HOLY LANDS THE FOUNDATION OF THE COMING BEE.

The question now arises, Have we any distinct race or strain of bees which shall form the basis or starting point upon which we may build? I would most emphatically affirm that in the new race, the Holy Lands, we have (as far as can be known) the original race of bees or the fountain-head; and further, that in them we find a larger number of the necessary qualities than in any other race or strain extant. I will admit that I am using strong language, but my association with Mr. D. A. Jones, at his home in Beeton, Ont., during the season of 1881, and my subsequent experience with the new race, regarding queen-breeding, honey-gathering and wintering, fully warrant me in making the statements which I do. Mr. Jones is most thoroughly practical in conducting his experiments in the apiary. He has thoroughly tested these races and given an honest description of them; and, although he has never received adequate remuneration for his whole-souled and extended efforts, yet posterity will render him the credit due him for the boon which he has conferred upon apiculture.

No race of bees will fly more rapidly or farther (when necessary) in search of honey, than will the Holy Lands, nor are any more hardy than they, and while not as gentle as *some* of our Italians, yet, when carefully handled, I have experienced no difficulty in this wise.

As honey-gatherers and breeders, they are not excelled, and woe to the luckless robber bees which attempt to invade their homes and fall into their hands.

They have been known to fly (in Palestine) six and one-half miles to obtain pasturage. I consider them a beautiful race of bees, although their type of beauty differs from that of any other race.

I have used bees for queen-rearing from nearly every race and many of the crosses, and find that the Holy Lands have no superior for this purpose.

At Beeton, in company with Mr. Jones, I have witnessed large numbers of Holy Land bees at work on the red clover (*large heads*) when there was a plentiful flow of white clover honey, and have known them to work on forage, in large numbers, farther from the apiary than the other races.

Mr. Jones stated to me that his Holy Land and Cyprian bees and their crosses stored from ten to twenty pounds more honey in the fall of 1881, and went into winter quarters in better condition than the other races.

These, with many other experiences and facts connected with the Holy Lands, fully support me in stating that in them we have the foundation on which to build the *Apis Americana*.

HOW SHALL WE DEVELOP THE COMING BEE?

The question now arises, What course shall we adopt in order to develop properly the coming bee? As previously stated, I do not consider that the Holy Lands (of to-day) nor any other race or strain of bees possesses all the necessary characteristics. While we find many desirable qualities in the Italian, German, Cyprian, and Holy Land bees, yet with Mr. Heddon and many other prominent, practical and successful bee-masters, I maintain that only a careful selection of those having desirable qualities from any or all the races, and the happy combination of all these qualities by careful breeding, will ever develop the long-hoped-for race.

There was never a greater curse pronounced upon
the future prosperity of apiculture than the cheap
queen traffic, and many long years will elapse ere we
can eradicate the direful effects produced by it; and I
deem it the duty of every intelligent bee-master, having
the interests and success of apiculture at heart, to
protest strongly against any measures which tend to
lower the dignity of this branch of industry, and I am
pleased to know that a majority of the most noted
and successful apiarists join hands with me in the
fact that the sooner the bee-keeping fraternity realize
the importance of this matter and patronize only
those who pay strict attention to careful selection
and breeding, the sooner we shall realize our expectations regarding the coming bee.

I look for the *best* results from a cross between
the young queens of our best *American Italians* and
Holy Land drones; with perhaps a mixture of German blood.

The largest and best working bees that I ever saw
were the progeny of a cross between the young
queens from an Italian mother, and Holy Land
drones, which I bred while with Mr. Jones. I intend
to make this matter a subject of careful experimenting
and study, to hasten, if possible, the time when we
shall have solved this question.

I will not attempt to describe the *modus operandi*
of the necessary selection and breeding, leaving this
to the author of this work, my teacher, the one to
whom I am most indebted for a practical knowledge
of scientific queen-rearing and to whose credit may

justly be said that "while our beloved Quinby and Langstroth were the fathers of practical apiculture in America," Mr. Alley has completed the work by introducing and establishing the only scientific and practical method of rearing queen bees.

In conclusion, I would state that I have spoken plainly in this matter because I have the best interests of apiculture at heart, and feel deeply the importance of a more thorough knowledge of this matter, and consider that when more strict attention is paid to rearing better (*not cheaper*) queens, we shall have made a grand move in the right direction; further, that this is the only road which leads to ultimate success, and would again urge my bee-keeping friends to study to select only the best, and pay careful attention to breeding if they wish to aid in producing the coming bee, or the *Apis Americana*.

Salem, Mass., Mar. 24, 1883.

GENERAL REMARKS.

THE OBSERVATORY HIVE A SOURCE OF PLEASURE AND PROFIT.

APICULTURE presents to the student of nature one vast and never-failing field of hidden treasure, surpassingly beautiful in its grandeur, while hours of pleasure and instruction may be profitably spent in searching it out. It has been said that, " the *apis* (or bee) is the last and most wonderful of the insect creation, as was man of the animal creation." Be this as it may, the bee in its construction of comb puts to shame the most advanced knowledge of mechanism, and its instinct amounts almost to reason. He who loves the study of the beautiful and grand in the Creator's handiwork cannot fail to find pleasure in studying the bees. Their form of government is monarchical and yet democratic; as, when the bees deem it conducive to the welfare of the colony that the queen be superseded, they commence at once to provide another, and seemingly with almost human affection; quite frequently they permit the old queen to live until her royal daughter assumes her position as mother bee, and then she is destroyed.

To make a colony queenless and then watch and study their movements and actions, through the various stages of development, while replacing their lost

queen, is a source of unceasing pleasure and grati-
fication. In fact, those who enter into the study of
apiculture, either for profit or pleasure, will find
themselves well paid for their efforts. One of the
best aids to this study is the observatory hive. By
its use we may study and experiment with the bees
without ·danger from their stings, and I know of
nothing used in connection with the apiary which
affords so much of genuine amusement and instruction to the student, family or friends, as does such a
hive. It is an endless source of amusement and
pleasure to the children to watch the queen and bees
as they move about on the comb, bring in and deposit
the honey, or their return from the fields with their
leg baskets filled with pollen. I have seen thousands
of people at our county fair (where I generally exhibit one of these hives) flock around and ply me
with innumerable questions about bees and their
habits : indeed, such exhibits generally attract the
notice of the larger portion of those who attend.

It may be placed in the office or parlor, and so arranged that the bees can pass out through the window,
without defacing anything, and really form a pleasing
addition to the furniture of the room. It may be
constructed and ornamented to suit the taste. All
the operations of the hive may be witnessed at any
hour of the day, and unless the glass is kept uncov-
ered too long at a time, the bees will not be disturbed
in the least. I feel confident that there are many
interesting and essential features of apiculture, which
are practically unknown to many bee-keepers (even

though they read the journals carefully and constantly), with which they may become familiar in the use of an observatory hive. For instance, the *modus operandi* of the bee in removing the pollen from its legs, the action of the bee when just made queenless, and many others. By having one of these hives at hand, not only these, but every operation of a full colony, except the storing of surplus honey in the caps, may be witnessed. Again, by its use, the first lessons in queen-rearing may be taken. When first made queenless the bees give evidence of the loss by running about the hive in a distracted manner, giving utterance to a loud and mournful hum or buzz, showing at once that the mother is gone. In the course of a few hours the bees will commence to construct cells in favorable localities near the centre, edge, or bottom of the comb. In a few hours several will be started, and more or less for two or three days after the queen is removed. In the course of three or four days the first cells will be sealed, and in eight days more the first queen will emerge. If the weather is favorable, she will be fertilized when she is five days old, and begin to deposit eggs in the course of forty-eight hours.

These are only a few of the many interesting operations which may be witnessed from day to day with one of these hives. If the apiarist has children who take an interest in such study, there is no possible way in which he can teach them more of apiculture, than by this. They soon become as familiar with the workings and operations of the bees as though they were lending a helping hand in the apiary.

SHIPPING BEES BY MAIL.

The practice of shipping bees by mail has been carried on for twenty years, and yet this part of the business has not yet been perfected and the results are still unsatisfactory. Experiments conducted in the season of 1882, however, were not only more successful than those of previous years, but gave results as nearly perfect as we may expect under the circumstances.

I. R. Good, of Nappanee, Ind., has devised a food composed of sugar and honey that has given good results, and yet it is somewhat imperfect, owing to the fact that when the bees have extracted the honey from it, the dry sugar crumbles, and is scattered about the cage, either stopping the ventilation or sifting through the wire-cloth into the mail-pouch; but, perhaps, after further experimenting, this difficulty may be obviated. The only advantage or benefit of the sugar is, that it retains the honey mixed with it, and in view of this is it not better to use an absorbent to retain the honey which is not subject to this imperfection?

During the season of 1882, I experienced just as good results from using sponges, properly filled with honey, in my shipping cages as did Mr. Good with his food; and not one queen died on account of a lack of food or manner of shipping.

The mail-pouches are handled very roughly, and quite often are left at some way-station and subjected to the scorching rays of the sun when the glass stands from 90° to 100° in the shade; again the weather may be quite warm at the shipping point, but before the bees reach their destination, they may meet a "cold wave," the temperature lowering to the freezing point and perhaps several degrees below. This, however, seldom occurs until after the middle of September. The cages are sometimes tampered with, the bees occasionally die from no apparent cause, and in view of this it is unreasonable to expect that every queen shipped will reach its destination in safety.

The cages should be neat, durable in their construction, and need not be bulky or heavy. There should be no tin or glass connected with them, as glass is unmailable, and tin is too cold, chilling the bees, and being so smooth that it offers no foothold, thus causing them to be thrown about the cages more. Ventilating holes one-fourth of an inch in diameter answer much better than larger ones, as they supply all the air necessary, and do not offer such tempting inducements to the inquisitive post-office clerks to experiment with them as they pass through the mails.

The cage that I use is three inches long, one and seven-eighths inches wide, and three-fourths of an inch thick. A hole one and a half inches in diameter is made through it, forming an apartment for the bees, and another hole an inch in diameter is made for the sponge as in fig. 15.

When but one queen is to be shipped, cut a piece of thin wood, the exact size of the cage, and one-eighth of an inch thick, and nail it on one side of the block;

then cut a piece of wire-cloth (twelve mesh to the inch), a little smaller than the cage, and after putting in the sponge nail it on the other side, using six (three ounce) tacks, leaving one corner unnailed until the bees and queen have been put in.

To put the queen and bees into the cage take it in the left hand with the wire-cloth facing you, and the lower end (or that containing the sponge) towards the wrist; the upper right hand corner of the wire-cloth which has been left unnailed should be

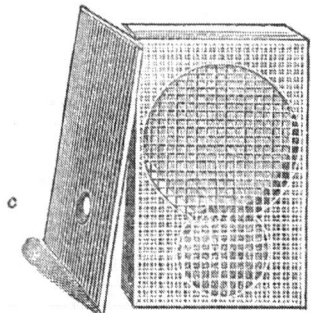

FIG. 15. Shipping Cage.

held open with the thumb and finger; next put in from six to ten bees[1], taking them by the wings and selecting the smallest bees that you can find, but not the very youngest nor those that are just ready to take a cleansing flight. After this has been done, take the queen carefully by the wings and put her into the cage, then slowly let the corner of the wire-cloth back into place, so as not to pinch the heads of

---

[1] When shipping queens that have been taken from the queen-nursery, or in any instance where you cannot use bees from the same colonies from which the queens were taken, take them from those that have been queenless three days or more.

the queen or bees, and tack it down. Now take the cap (c) which has the ventilating hole in it, and nail it over the wire-cloth, using four three-quarters inch wire or cigar-box nails, putting one in each corner.

Bees thus prepared will go safe when they are not more than seven or eight days on the road, but when they are to be from ten to twelve days in transit, make the cages nearly twice as thick (but no larger otherwise), as this gives room for a larger sponge and double the amount of honey.

When two queens are to be shipped to one address, cover both sides of the cages with wire-cloth, nail the ventilating cap on one side of each, and then place the face (or wire-cloth covered) sides together, reversing the sponge ends, so that the bees in either cage may take food from the sponge in the other.

When sending three or more queens the middle cages will not need any caps, but the sponges should be reversed as in the former instance, and the bees will be sufficiently ventilated through the outside cages. It would not be advisable to ship more than six queens in any one lot. When the cages are ready to pack, place them in a vice, and after pressing them together tie a strong string around them.

The food in the sponges is composed of four parts honey and one part water, and in filling them the surplus food should be pressed out in order that it may not run out and daub the bees or mail. The small hole for the sponge should be lined or coated with hot bees-wax and rosin. This, however, is not necessary unless the bees are to be four or more days in transit. A label should be pasted on one side of the cage, with the address plainly written upon it. By not adopting this method I lost seventy-five queens in the season of 1882.

In conclusion, I would again caution queen-dealers to be very careful, both in preparing and shipping bees by mail. The cages must be strong and durable or they may be crushed, and no food should be used which is liable to injure the mails: such as loose honey, honey in the comb, or sugar food that will crumble and sift out of the cages. The wire-cloth should be covered in such a way that the bees will be protected from injury at the hands of inquisitive post-office clerks, and the clerks be protected from the stings of the bees (I have received bees in cages when the wire-cloth was unprotected) ; for, perchance, unless this is done some clerk may pinch the protuding foot of some luckless bee and receive a "stinging rebuke." This and other careless neglect may cause complaint and result in the suspension of the important and invaluable privilege of shipping bees by mail.

A NEW WAY TO TEST QUEENS.

The hives containing queens to be tested should be numbered and a correct record kept of the queens, their origin, age, the time when they commenced to lay and all the necessary particulars concerning them. Now, if there is not room for all the maturing cells in the nuclei, queens to be tested may be removed to the nursery after they have been laying for one week or more. The cages must be well supplied with food and the numbers of the queens marked upon them. The nursery should then be placed in the brood-chamber of some full colony.

Queens may be kept in this way until their worker progeny has hatched. If the brood left by the queens was put in the combs solid, and the young bees hatch out strong, vigorous, and show unmistakable signs of purity, the queens may then be shipped at once. The colonies thus made queenless may be either supplied with cells, or, if these are not at hand, with young queens introduced by the method previously explained.

CLIPPING QUEENS' WINGS AND SWARMING.

I have never practised clipping queens' wings as I consider it not only useless but also very injudicious. I am not prepared to say that it is any direct injury to the queens, but I know that it does not prevent swarming; instead, it surely makes matters worse when this does take place. When bees are prepared to swarm they will do so notwithstanding the queen has but one wing, and if the hive is on a stand one foot or more above the ground, the queen may be lost, and if she is not lost but returns to the hive, the swarm will issue again and continue so to do unless, in the meantime, the old queen is destroyed. When the young queen emerges from the cell, the swarm will then issue if the weather is favorable.

11

HOW TO PREVENT SWARMING.

If the apiarist does not wish to increase his number of colonies, he may unite two of the swarms in one, giving them sufficient room in which to store their honey. The only way that I know to prevent swarming is by interchanging the queens, first removing them and keeping the colonies queenless for three days; then again introduce the queens by directions given on page 26. The cells which have been started will be destroyed. This has the effect of diminishing the swarming fever.

Swarming, however, cannot be prevented without considerable trouble, and I know of no practical way to prevent it altogether. When they have swarmed, it is about as well to let the bees have their own way in this matter, and control a too rapid increase by doubling up the colonies as they issue.

HANDLING NEW SWARMS; HIVING, ETC.

When a swarm has issued, it should be allowed to settle before being disturbed, and when the bees have become quiet, it is a good plan to sprinkle them with water, as it causes the bees to cluster more closely and also checks their disposition to fly or sting, as they at times seem inclined to do. I find that the "Whitman's Fountain Pump," fig. 16, is just the thing for this purpose.

Bees sometimes settle upon the limb of a tree, a number of feet from the ground, making it quite difficult to reach them, but by the use of this pump, they may be easily sprinkled, thus completely subduing them. Again, if the bees do not seem disposed to alight, the use of the fountain pump will soon cause them to rush "pell-mell" for some place on which to settle.

FIG. 16. Whitman's fountain pump.

I generally use a bushel basket in hiving bees when they cluster in a convenient position ; or, if hemlock-saplings can conveniently be obtained, cut a light one and trim off all the branches excepting those at the end. Now, if the swarm has clustered, work the brush end of the sapling in among them and then slightly jar the limb on which the bees are clustered; continue to do this until there is quite a cluster of them on the sapling, then with a sudden jar shake the remaining bees from the limb, standing one end of the sapling on the ground to support it and carefully swinging it away from the tree and holding it there until nearly all of the bees have settled on it. They may then be carried and shaken down in front of a hive prepared to receive them.

Place the hive upon the stand, grass, or in any convenient and handy position; then, if the basket is used, shake the bees (by a sudden jar) into it, and turn them down in front of the hive, paying no attention to those which take wing as they will soon return and enter with the rest, but should some of the bees still persist in clustering on the limb, jar them off until they discover and enter the hive; after this has been done and the bees have all entered, place them upon the stand which they are to occupy.

THE USE OF TOBACCO.

Perhaps my friends may conclude that I have recommended tobacco for introducing queens, because I am addicted to the habit myself. I would here state that I never have used the *filthy weed* other than for the purposes stated, and although I have used tobacco in my apiaries for twenty-five years, yet I never have become so accustomed to it that the taste, and even the smell of it are not obnoxious to me.

For fumigating bees, I use a smoker constructed as follows : a tin tube six inches long, and seven-eighths of an inch in diameter, with a wooden stopper in each end. The one through which the smoke is blown has a small tin tube running through it projecting about one inch beyond the end, so that the smoke can be directed to any particular point or part of the hive. The other forms a mouthpiece and has a small hole bored through it (lengthwise of course). The smoker is held between the teeth leaving the hands at liberty. To light it, remove the mouthpiece, place the small tube in the mouth and proceed the same as with a pipe or cigar. I have had occasion to use this pipe as many as fifty times in one day, but this was before the bellows smoker came into use.

IN-AND-IN BREEDING.

I do not think that sufficient attention has been given this subject by the writers in the various bee journals ; it might well replace many subjects of minor importance.

In my experience as a queen-dealer, I have always kept in mind the importance of proper selection and careful breeding as connected with the improvement of the bee. The drone as well as the queen should be up to the standard in points of excellence in every respect.

In rearing queens more attention should be paid to the selection of the drone than of the queen, as the former exerts the greater influence in transmitting qualities. I do not wish my readers to think that I would undervalue the influence of the queen-bee in this matter; on the contrary, the best possible selection should be made of both sexes, and in-and-in breeding should be avoided as much as possible, although I am not sure that its results would be felt in a large apiary for several years.

In some counties where bees are kept in considerable numbers, I have learned that nearly all of them originated in some one prominent apiary, and the bees to all appearances continued hardy and vigorous.

The Cyprian bees have been confined for ages upon Cyprus Island, and who will say that *they* are not both hardy and "*vigorous*."

While for a time at least no apparent evils may result from in-and-in breeding, yet I would again advise a careful introduction of good-blooded stock into our apiaries, and also that in-and-in breeding should be carefully avoided by every possible means.

BEE PASTURAGE.

It has been truthfully stated that the demand for first-class honey far exceeds the supply, and even though the millions of pounds now produced be doubled or trebled, yet the demand will proportionately increase. With a better knowledge of the best methods of producing and marketing honey, we are reducing the attending expenses and increasing its production; and although (as has been the case with all new articles of food) we find many obstacles to surmount in creating a market for our honey, yet we may feel gratified to know that it is fast becoming a staple article of commerce and is gradually taking the place of many of the syrups.

One great difficulty, however, stands in our way and demands our careful consideration and attention. I refer to pasturage for our bees. From the first vivating touch of the warm sunshine of spring, until the landscape is robed in white for a winter's rest, the supply of forage from which our surplus honey is gathered, is irregular; or rather there are many "dead points" or honey dearths.

The vast and luxuriant forests of basswood, which in the past have proven of so much value to the apiarist, yielding bountiful supplies of nectar, are fast giving way beneath the woodman's axe, and ere long will disappear, thus depriving us of one great source of revenue.

These facts have led many of our most prominent and thoughtful apiarists to give this subject much careful study and attention, and they have invariably come to the conclusion that it is not only advantageous, but remunerative, to sow or plant for bee-pasturage. The thousands of acres of now almost worthless land, in various portions of our country, if utilized for this purpose, would return a handsome revenue and thus add to our individual and national prosperity.

HONEY-PRODUCING TREES.

The planting of honey-producing shade trees such as the sycamore maple (*Acer pseudo-platanus*), horse-chestnut (*Æsculus hippocastanum*), and basswood (*Tilia Americana*), would not only beautify our landscape and form good wind-breaks, but also give (to the coming generation at least) an additional source of revenue.

It is becoming imperative that we give this matter thoughtful study and test some of the best honey plants, choosing those best adapted to the localities in which we reside.

THE BASSWOOD.

Among the trees, the basswood stands first in value, and in fact it has no equal. It affords a luxuriant shade, and it is really a pleasure to watch the thousands of busy workers, flitting to and fro, sipping the precious nectar from the well-filled flowers, and to listen to the sweet music of their busy hum.

THE HORSE-CHESTNUT.

The horse-chestnut is quite a prominent shade-tree here in New England, grows fast, and yields a good quality of honey, the flow lasting for ten days at least. The young trees may be procured from any reliable nurseryman.

BOKHARA.

Among the honey-producing flowers the Bokhara or sweet clover (*Melilotus alba*) (the white variety) ranks first. It is almost impossible to find a locality in which it will not thrive, from the dry, barren sand hills to the low bog-holes and marshes.

Every apiarist should convert himself into an individual seed-drill, and carry about in his pocket a quantity of the seed, scattering it along the roadside, in old waste pastures and in fact, in every out-of-the-way place which is now running to waste.

When sowing the sweet clover alone, use five pounds to the acre. It may be sown broadcast in the spring when the ground is wet and then light-harrowed in.

This plant will prove a great boon, especially to New England, and I look forward to the time when many portions of our country, which now hardly yield enough to pay the taxes, may prove remunerative from the production of honey. Many instances might be cited to substantiate this, but every one who has tried or will try this clover seed may learn this to be a fact.

ALSYKE CLOVER.

Next in value comes the alsyke clover (*Trifolium hybridum*) a native of Sweden, which is both hardy and prolific, and produces a fine quality of hay. The honey produced from it is unequalled for quality and flavor by any other that I have ever seen. When sown alone it will take four pounds to the acre, but it is better to mix it with timothy (*Phleum pratense*) or the common red clover (*Trifolium pratense*) or both, using the usual quantities of the latter and two pounds of the alsyke.

The timothy and red clover help support the alsyke, and there is less danger of its growing too thick and lodging.

WHITE ALDER.

The white alder (*Clethra alnifolia*) is another fine and valuable honey plant. It is a native of Massachusetts and is most generally found growing between swamp and upland, spreading freely into ploughed lands bordering on its natural thickets. As an ornamental plant it is a neat, upright growing shrub. Its leaves are light green, flowers all pure white, in spikes three to six inches long. It blossoms late and through a long season. This plant is now attracting much attention as a forage for the honey bee. It is practical to plant for this purpose by the acre. It transplants safely, and is propagated very easily by suckers and layers; we plant it here until May 15, and from October until December. It leaves out late in spring and blossoms on plants one to eight feet high, according to age and vigor of growth. A strong plant in vigorous soil would make 100 plants in three years, and the planter of 1,000 can extend its culture to acres. The honey from this plant is almost colorless, heavy, rich and of a fine flavor.

I am well acquainted with this honey plant and take great pleasure in pronouncing it one of the best known for general use.

There are a number of good honey-producing plants, but I have enumerated only those best adapted to general cultivation; and of those which I have mentioned, I would strongly recommend the bokhara or sweet clover, as it can be sown with little trouble, and then resows itself, and is not easily killed out; also it continues to yield honey, during favorable weather, all summer and until the late frosts kill it, the bees often leaving the white clover and giving the sweet clover the preference.

I would advise every reader to procure some of the seed and give it a thorough trial, remembering that it starts best when it has been subjected to at least one good frost.

DESCRIPTION OF FIGURES.

DOUBLE-WALLED HIVES.

HIVES of this kind are coming into general use and are growing rapidly into favor. A few years since I devised a double-walled hive on a different principle than any then in use.

Most of those that have been used for the past twenty years have been too heavy and clumsy, and so constructed that the bottom-board and outer case were nailed together. An improvement on such a style was demanded, and in view of this I constructed one after the following plan:

The outer cases, bottom-board, brood-chamber, and in fact each particular part of the hive, are independent of each other, nor is any part of it secured to the bottom-board with nails. The brood-chamber rests on the bottom-board (B) and is held in place by the strips (C and D) which are nailed to it as seen in fig. 17.

The comb-honey rack (G), which contains the boxes (O), rests on the top of the brood-chamber, no honey-board being used. The outer cases (J, L, fig. 17) rest on the bottom-board and cover all, thus forming a complete protection against the sun, rain and severe winter weather.

I claim that double-walled hives are indispensable in our northern sections especially; and just as necessary in summer as in winter, as the bees need to be protected and shaded from the burning rays of the summer's sun, as well as from the cold and chilling blasts of winter. I further claim that there is no need of filling the space between the outer case and brood-chamber with any material whatever, a chaff cushion over the bees being all that is needed.

An ice-house, however thick the walls might be, would not preserve the ice in them, if a space is left open for the air to enter; neither will a colony of bees be thoroughly protected from the cold or the changes in the weather by thick walls when the entrance is left open in the usual way.

The hive is so constructed that a comb-honey-rack holding twenty-one two-pound boxes will just cover the brood-chamber, and two of these racks can be used on one hive by placing one above the other, thus giving room for forty-two boxes; a second story may be placed on the top of the brood-nest, thus giving room for an extra set of frames, for use in extracting or obtaining comb-honey after Mr. House's plan. I shall adapt my hives for either purpose.

This hive requires the least labor in its construction of any in

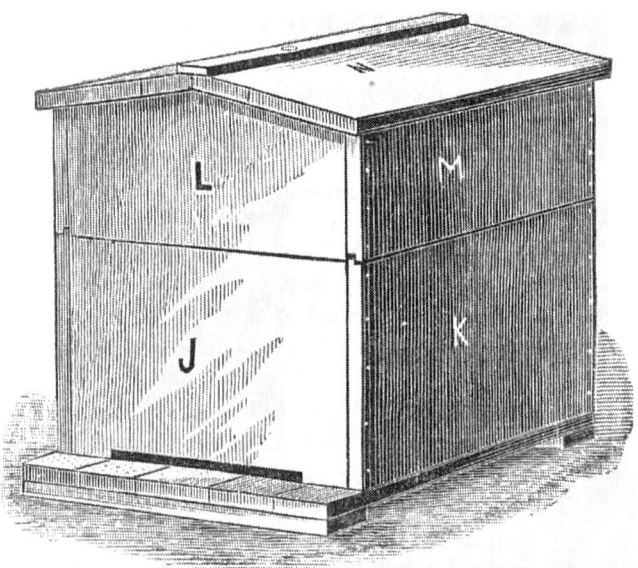

FIG. 17. Outer case of double-walled hive.

use. I use the standard Langstroth frames with the exception that the top-bar is one and one-eighth inches instead of seven-eighths of an inch wide. I have used nine frames to a hive for four years and find this number preferable to ten frames.

For the convenience of my readers I would give the following directions for its construction.

The bottom-board (B) 27½ in. × 18⅝ in. × ⅞ in. may be made of two or more pieces. The clamps (A) 18⅝ in. × 1½ in. × ⅞ in. should be nailed on the under side. The strips (D) 22¼ in. × ⅞ in. × ⅞ in. are nailed to the bottom-board fifteen-sixteenths of an inch from the outer edge, and seven-eighths of an inch on from the back end, and they should be so nailed that the outer case will slip down over them easily, as they are intended to keep it in place as well as the brood-chamber. The strips (C) 16⅞ in. ×1½ in. × ⅜ in. are nailed across the strips (D) flush with the ends, thus holding the brood-chamber in place, and forming, in connection with an opening in the front of the brood-chamber, and a like opening in the front of the outer case, an entrance for the bees.

The brood-chamber is composed of six pieces: the sides (H) are 19⅞ in. × 10 in. × ½ in.; the front (E) is 14⅛ in. × 8⅜ in. ×

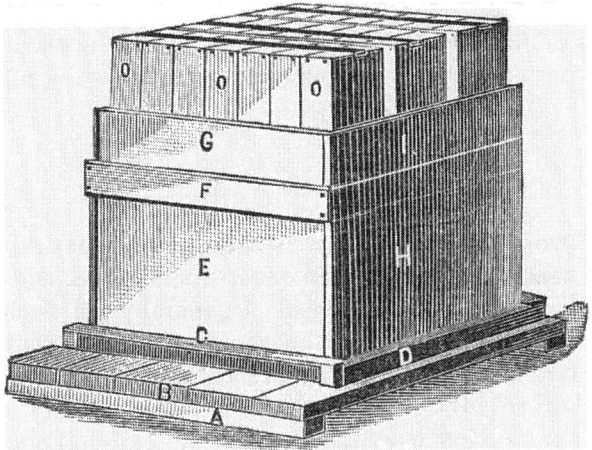

FIG. 18. Inner case of double-walled hive.

⅞ in., and when nailed is set down from the top edge of the brood-chamber nine-sixteenths of an inch. The back end 14¼ in. × 9⁹⁄₁₆ in. × ⅞ in. is also set down nine-sixteenths of an inch, thus forming a shoulder on which the frames may hang. Strips (F) 15⁷⁄₁₆ in. × 1⅝ in. × ½ in. are nailed flush with the top edge of the brood-chamber and complete the rabbet for the frames. I have not as yet completed the improvements intended regarding the surplus arrangement as suggested by Mr. House's article.

The outer case, bottom section, consists of four pieces: the sides (K) being 24 in. × 11½ in. × ⅞ in., and the end pieces (J) 16⅞ in. × 11½ in × ⅞ in., one of them having an entrance cut in it 14 in. × ⅜ in.

The top section, side pieces (M) 24 in. × 6⅞ in. × ⅞ in., the end pieces (L) 16⅝ in. × 10 in. × ⅞ in., are bevelled from the centre to the ends, so that they will correspond with the ends of the side pieces, thus giving the roof a pitch of about one and one-half inches. The roof boards (N) are 26 in. × 10 in. × ⅞ in., and the piece (O) to cover the joint is 26 in. × 2 in. × ½ in.

In nailing up the hives the longer pieces are nailed to the shorter ones, using eight-penny finish nails for the heavier parts, and smaller nails for the rest.

The edges of the sides (M and K) of the outer case should be halved together to keep out the rain, etc.; groove on $\frac{1}{16}$ in. and $\frac{4}{16}$ in. deep.

QUEEN-REARING HIVE.

This is a two-story hive, taking twenty-four nucleus combs, such as are used in the miniature or fertilizing hives, and will accommodate a large colony of bees. I generally use them for queen-rearing, and as they take the same size frame that the small hives do, I find them much more convenient for the purpose than standard hives.

The prepared strips containing eggs for cell-building are placed in frames as shown in fig. 6, from two to four of these being used according to the size of the colony, which are placed in the upper story. I think it just as well to place them in the lower story, but it is not so easy to get at them when you desire to examine the combs to ascertain the number of cells being built and for other purposes.

By using twenty-four combs we give the bees more room for storing honey, and as the colonies in the miniature or fertilizing hives do not always store enough for their own use, the full combs may be exchanged for empty ones, thus saving a vast amount of time and trouble in supplying such colonies with food.

If a colony put in one of these hives be very large, a third story is supplied with frames filled with foundation, or perhaps a case filled with sections should be given them. Old combs may be placed in the third story and the honey extracted from them as well as from those in the second story. As stated elsewhere, a colony deprived of its queen and brood will store honey rapidly, there being no brood in the hive to care for and no hatching bees to consume the honey as it is gathered.

A colony in one of these hives will care for a large number of sealed queen-cells when necessary. When I have two hundred sealed cells and wish to use the bees which have just completed them, to form nuclei or for other purposes, I "double up" by placing all the cells in one hive, keeping of course a correct record of each separate lot, otherwise many queens would be lost. In rearing queens in these hives I proceed in the following manner: one hive is kept queenless for the purpose of caring for cells, and when one colony has sealed one lot of cells and I need the hive in which to start others, the cells are placed in the hive kept for that purpose; all the bees are then removed from a

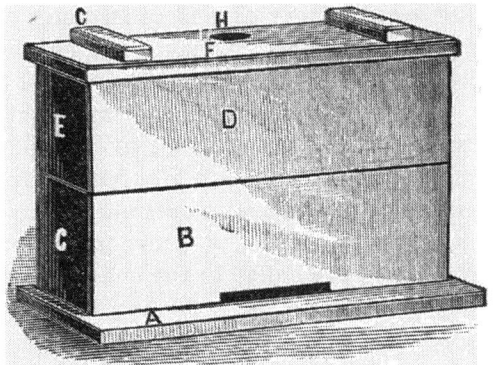

Fig. 19. Queen-rearing hive.

full colony, placed in a swarming box, and the bees which have just completed the cells are put into the hive from which the others were taken, giving to them the old queen.

If you should desire to form nuclei of the bees and combs after the cells are sealed, proceed thus: remove all the cells to the hive kept for that purpose, or to a fertilizing hive, being careful that the colony has enough bees to protect the cells; now, if you have the fertilizing hives at hand, the bees and combs can be transferred to them in a few moments. First, prepare the hives by placing the screens over the entrance; then place the combs with adhering bees into each hive, set them away in a cool place for three days, and at night on the third day, give to each a cell or queen (virgin queens can be

12

safely introduced to such colonies) and one frame of brood. After dark remove the screens and allow the bees to fly on the following morning.

To manage these hives properly one must have plenty of small combs for them. I would advise "breaking up" colonies in box-hives to obtain these, as it would be much better and cheaper than to cut up nice straight combs.

The following measurements should be taken into account in the construction of these hives:

The bottom-board (A) is $21\frac{1}{2}$ in. $\times$ $11\frac{1}{4}$ in. $\times$ $\frac{7}{8}$ in. The front and back (B) are 20 in. $\times$ 6 in. $\times$ $\frac{7}{8}$ in. with a rabbet on the inside top edges, $\frac{7}{16}$ in. $\times$ $\frac{1}{2}$ in. on which the frames may hang, and also rabbeted on each inside end, $\frac{7}{8}$ in. $\times$ $\frac{7}{16}$ in. to form shoulders for the ends of the hive ends (C) which are 7 in. $\times$ 6 in. $\times$ $\frac{7}{8}$ in.

The upper or second story (D) is constructed the same as the lower or first story, excepting that it has no bottom-board. The top or cover (F) which is made of one piece is $20\frac{1}{2}$ in. $\times$ $8\frac{3}{4}$ in. $\times$ $\frac{7}{8}$ in., with two cleats (G) nailed across it to prevent warping, H representing the aperture over which the feeder may rest. Make an entrance 7 in. $\times$ $\frac{3}{8}$ in. in the lower story.

The frames are composed of four pieces: top-bars $6\frac{7}{8}$ in. $\times$ $\frac{7}{8}$ in. $\times$ $\frac{1}{4}$ in; bottom-bars $5\frac{3}{8}$ in. $\times$ $\frac{7}{8}$ in. $\times$ $\frac{1}{4}$ in.; ends $4\frac{7}{8}$ in. $\times$ $\frac{7}{8}$ in. $\times$ $\frac{1}{4}$ in. The bottom-bars are nailed to the ends.

MINIATURE OR FERTILIZING HIVES. (Fig. 1, page 4).

The terms miniature and fertilizing hives are synonymous or, in other words, these hives are alike in construction, excepting that while the miniature hive takes five frames, the other takes but three or four, and I have only used these separate terms in order to designate them. The nucleus hive, to contain the breeding queen, should take five frames for reasons given on page 5.

The bottom-board (a) of the fertilizing hive is $8\frac{3}{8}$ in. square with the grain of the wood running from side to side; the top or cover is the same but has two clamps nailed on the upper side to prevent it from warping. The front and rear pieces are $6\frac{1}{4}$ in. $\times$ 6 in. and are rabbeted down from the upper edge $\frac{1}{4}$ in. and in from the inner edge $\frac{3}{8}$ in. (the same as are the queen-rearing hives). The end pieces b are 6 in. $\times$ $7\frac{3}{4}$ in.

The miniature hives for the breeding queens are constructed the same as the fertilizing hives excepting that the pieces for the front and back are 8 in. long instead of 6¼ in. as in the former instance.

CONE FEEDER. (Fig. 13, page 37.)

About twenty years ago I obtained my first feeder of this kind from a man in New Hampshire, and have used such ever since.

The principle is atmospheric; this feeder holds about four ounces (erroneously stated one ounce, on p. 37), a sufficient quantity to keep a colony consisting of about one pint of bees at work twelve hours, in removing and storing it. You will notice that it has a tin collar (a), at the small end; b representing the collar separate from the feeder. After filling the feeders, place a thin piece of cloth over the end and then put on the collar to keep it in place, and to prevent the food from running out. Such feeders are an almost indispensable article for feeding small nuclei. The reader will notice how they are used on the hives. The one inch hole in the cover of the small hive, described on page 4, is intended for the insertion of this feeder; and when the collar has been properly put on and the feeder placed in position (as it should be) the rain will not leak in around the joint. I generally bore the hole in the cover near the front, and elevate the rear end of the hive a little, so that should any water chance to leak in it will run out at the entrance and not back into the hive.

The piece of cloth placed over the end of the feeder should not project beyond the collar as it would operate the same as a lampwick and be wet (with the liquid-food) all the time and offer too great a temptation to robber bees. Again, the collar must be adjusted carefully so that no air can enter the feeder, else the syrup would run out, and induce robbing. However, when simple sugar-syrup is used there is not much danger of robbing.

LOCKE'S FEEDER.

Fig. 20 represents a feeder devised for the purpose of winter feeding and spring stimulating with either water or syrup, and

I take great pleasure in pronouncing it unequalled for this purpose. By its use I have saved fifteen colonies from starving, this spring, as they had eaten out the honey from the centre of the hive and were unable to change position on account of the unusually extreme cold weather. I would recommend it to the notice of my readers.

It consists of an entirely new and original combination of principles. It holds about one pint of syrup which is supplied from the top of the feeder without disturbing either the bees or feeder and taken from a sponge at the bottom. After it is filled and the screw-top adjusted, it becomes an atmospheric feeder. The under side is covered with cloth secured with bees-wax and rosin rendering it impervious to moisture. This cloth forms a non-conductor of heat and permits the bees to cluster

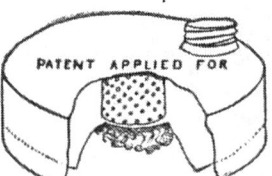

FIG. 20. Locke's feeder.

under it and around the sponge without becoming chilled. This is essential, as we often need to stimulate bees before we can examine the colonies to contract the brood-chamber, and quite necessary when we have severe cold weather in March, as the bees (especially those wintered on summer stands) eat out the honey in the centre of the hives and perish by starvation rather than leave the brood which may have been started, or break the cluster. With this feeder the bees need not leave the cluster to reach the food. To use it, cut an oblong hole through the cloth-cover nearly as long as the diameter of the feeder and about one inch wide, crosswise the frames; this will give the bees a chance to reach the food from five or six of the combs. After placing the feeder in position over the hole, remove the screw-top, fill the feeder and then adjust the screw-top as speedily as possible. Then cover it with the chaff cushion. The feeder may be placed upon the hive in the fall and remain all winter without detriment to the bees.

For years there has been a demand for such a feeder and I consider this one sufficiently meritorious to warrant a description for the benefit of my readers.

THE SWARMING BOX.

The box alluded to on page 8 forms one of the handiest and most useful pieces of furniture in the apiary, and every apiarist should have one or more of them.

They may be used for many purposes other than to confine bees preparatory to cell-building, such as uniting and transporting bees, introducing queens, etc., which are mentioned in different parts of this book.

The main box is 20 in. long, 10¼ in. wide, and 10 in. deep, inside measure; the sides are 21¾ in. long, 10 in. wide, ⅞ in. thick and are nailed to the ends, after which the bottom is covered with wire-cloth twelve mesh to the inch, as coarser mesh would let the heads of the bees through and many would be killed; besides, there would be great danger of suffocation as the bees would nearly stop the ventilation.

Next nail two strips 12 in. long, 2 in. wide, and 1¼ in. thick across the bottom at the ends (and crosswise the box), flush with the outer edge; another piece ⅞ in. wide and ½ in. thick, across the centre of the bottom to keep the wire from sagging when bees are in it. Now nail some thin strips between these pieces to keep the wire-cloth in place.

The cover is made of five pieces; two of them are 1½ in. wide by 20 in. long, the other three pieces are 2½ in. wide and 8½ in. long. The longest pieces are nailed to the three short ones, one of the latter is nailed in the centre. It is now covered with wire-cloth and some thin strips are nailed around the edges of the wire to keep it in place.

A piece of heavy tin 12 in. long and 1 in. wide is nailed to the under side of one end, and one exactly corresponding with this is nailed on the top edge of the box to come under the one on the cover (the purpose of this has been stated on page 17).

QUEEN NURSERY. (Fig. 9, page 23.)

The queen nursery may properly be ranked among the essential and even indispensable fixtures of a complete apiary and no bee-keeper can afford to do without it; still more necessary is it if the apiarist is a queen-breeder. It will save its cost in one season during swarming time and must be resorted to (in the queen-yard) during a long spell of dull weather, or the worse than useless lamp-nursery system must be adopted temporarily. The latter method is so contrary to natural laws and so injurious in its effects that I consider it my duty to condemn it to say the least. I claim that the cheapest and simplest implements used in the apiary are the best, provided they are properly constructed and answer the purpose as well as the more expensive ones. The nursery which I am about to describe is a good and convenient one and answers the purpose well. I had never seen any other before I constructed this one, although I had read descriptions of them in the various bee journals.

The nursery is composed of a frame similar to a standard Langstroth frame and holds eighteen cages. The cages (fig. 10) are cut $2\frac{11}{16}$ in. long, $2\frac{1}{2}$ in. wide, and 1 in. thick. A $1\frac{1}{2}$ in. hole is made in about the centre. In the long edge of the blocks and $\frac{3}{8}$ of an inch from the end, is bored another hole $\frac{3}{4}$ in. in diameter, and 1 in. deep for the sponge containing the honey to feed the bees, and as this hole breaks into the large one, the bees can take the food from the sponge. Another hole is made (on the same edge) in which to insert the cell, which is held in place by pressing the wax on the upper end against the wood. I experience no trouble in doing this, but as some of my readers may find it difficult to fasten the cells in, in this way, I would advise running a nail through the upper edge of the cage and through the top of the cell.

These cages must be sawed very accurately in order that they may closely fit in the frames.

INTRODUCING CAGE.

I have for several years used such a cage as I am about to describe. I first found it necessary to adopt it when at one time I was called away from home to Italianize a lot of bees, and could spare but one day for the work; I therefore made a cage of this kind and it has proven so successful, I will try and describe it so that any one may make it, although I am aware that even

with the most plain and simple directions it may be difficult for
the reader to construct it properly.

Take a block of wood 3 in. long, 2 in. wide and $\frac{1}{2}$ in. thick, and
bore through it a $1\frac{1}{4}$ in. hole one-half inch from one end; then
take a knife and cut the slot or mortise (a) from the hole to
the end of the cage (or block), being careful not to cut out
more than enough to allow the bees to pass through after the
wire-cloth is fastened on. Now cover both sides with wire-
cloth (as seen in fig. 11); next cut the piece of tin (b) $1\frac{1}{2}$ in. long
and $\frac{1}{4}$ in. wide, and fasten it to one end of the cage by driving a
wire nail through the centre of it and into the cage.

This is adjustable and works on the principle of a button to
a door; and when it is turned crosswise (as shown in cut)
the cage will hang between the combs and thus will be held in
position and prevented from falling down between the combs.
This cage may be used in introducing both laying and virgin
queens. The queen should be put in through the mortise-hole,
which should then be filled (or plugged) with a mixture of
sugar and honey (Mr. Good's food answers this purpose well);
and in introducing, if the bees have been queenless three days,
the queens will be kindly received; but as this matter is fully
described on page 26 I will not further explain it.

NEW WAY OF HAVING CELLS BUILT. (Fig. 6, page 15).

This figure represents one of the small combs used in the
fertilizing hives, with a row of cells built by the bees, and as
this cut has not been sufficiently described, I will here explain it
more fully. It shows the manner of placing in position the prepared strip containing eggs for cell-building; and also shows the
cells after they are sealed. The cut shows an open space between
the cells; now the bees do not leave them in exactly that shape,
but connect them by a thin septum or partition of wax. I do not
wish it to be understood that they *join the cells one with another,*
as, if they did so, it would be impossible to separate them without
injury, but when the bees are gathering honey rapidly, they
will cover the queen-cells with wax often building cells upon
them nearly to the points, filling them with honey.

When this happens I usually shave off the honey clean, from both the cells and the strip upon which they are built, leaving them in the hive until the bees have cleaned the honey from them, otherwise the operation of separating the cells would be a sticky one. It is not necessary to shave off the honey until a few hours previous to separating and transferring them to the nursery or nuclei. It seems rather a delicate operation to place in position the "prepared strip." This is so when the comb containing the eggs is tender, but you should avoid as much as possible using new or tender comb for this purpose.

When standard frames are used, the prepared strip may be placed near the bottom of the comb by cutting out a small piece to make room for the cells and thus avoiding the necessity of destroying good combs.

GLEANINGS

—:IN:—

BEE CULTURE

TERMS, $1.00 PER YEAR.

GLEANINGS is a 52-page monthly, published the first of each month. A supplement, JUVENILE GLEANINGS, of 16 pages, is published the middle of each month, thus giving reports semi-monthly of the state of bee-culture in almost all regions of the globe where civilization extends.

A B C OF BEE CULTURE.

A book of 318 pages, and over 175 illustrations. This is kept standing in type, and corrected several times a year; thus when you buy it you are sure of a book that is up with the times. Price, in cloth, $1.25 ; in paper, $1.00.

APIARIAN IMPLEMENTS AND SUPPLIES,

Our customers now number something over **75,000,** and goods are shipped to all parts of the world. To keep pace with late improvements and new inventions, our price list is kept constantly standing in type, and new editions are printed in the busy season, frequently as often as once a month. A sample copy of GLEANINGS and a price list sent free on application.

A. I. ROOT,

MEDINA, . . OHIO.

THE BEE-KEEPER'S GUIDE;

OR,

MANUAL OF THE APIARY,

By A. J. COOK,

Professor of Entomology in the

State Agricultural College, Lansing, Michigan.

10th 1,000
350 Pages.   190 Beautiful Illustrations.

This is a new edition, the second full revision of Prof. Cook's Manual of the Apiary, enlarged and elegantly illustrated. The first edition of 3,000 copies was exhausted in about 18 months, the first 9,000 in six years — a sale unprecedented in the annals of bee-culture. This new work has been produced with great care, patient study and persistent research. It comprises a full delineation of the anatomy and physiology of the honey bee, illustrated with many costly wood engravings—the products of the Honey Bee; the races of bees; full descriptions of honey-producing plants, trees, shrubs, etc., splendidly illustrated—and last, though not least, detailed instructions for the various manipulations necessary in the apiary.

This work is a masterly production, and one that no bee-keeper, however limited his means, can afford to do without. It is fully "up with the times" on every conceivable subject that can interest the apiarist. It is not only instructive, but intensely interesting and thoroughly practical—*T. G. Newman.*

It is a credit to the author as well as the publisher. I have never yet met with a work either French or foreign, which I like so much.—L'ABBE DU BOIS, editor of *Bulletin d' Apiculteur*, France.

It appears to have cut the ground from under future book-makers.— *British Bee Journal.*

Messrs. A. J. Root, L. C. Root, and Thos. G. Newman, all say that it contains much not found in other books, and all the distinguished writers on bee-culture recommend it.

Sent on Receipt of Price $1.00.

A. J. COOK, Author and Publisher.

KING, KEITH & CO.,
IMPROVED HONEY EXTRACTOR.

No. 1. 4-FRAME MACHINE.

Five Styles of Two- and Four-frame Honey Extractors, extracting from *all frames in use*, and even small pieces without frames, *complete* in EVERY RESPECT, holding from 30 to 225 lbs. *below* the basket; also Wax-Extractors, Bee-Hives, Bees, Queens, Smokers, Foundation and Machines; Books on Bees, and in fact *every article* of use in the Apiary.

Send for our Illustrated Catalogue for 1883, *before purchasing.*

KING, KEITH & CO.,

Pub. of new *Bee-keepers' Text Book* and the 40 page *Bee-keepers' Magazine*,

14 Park Place, New York.

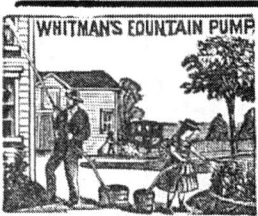

Comb-Foundation, Wholesale and Retail.

BRO. BEE-KEEPERS: I have ordered two new mills of Mrs. Dunham, and am ready to fill orders for any size and quantity of the finest Comb-Foundation.

Please send your orders early, secure your foundation and avoid delays in the busy season.

I will pay Thirty cents per lb. for pure clean Wax delivered at *our* Rail Road Station. The shipper must put his name on each package.

Those sending Wax to be worked up will have one-half the amount returned in foundation: the person sending the Wax to pay freight charges both ways.

If you will give me a trial I will endeavor to please you. All orders will receive prompt attention and no delays will occur which I can prevent.

Satisfaction Guaranteed or Money refunded.

SEND FOR SAMPLES.

PRICE LIST OF FOUNDATION.

HEAVY.	MEDIUM.	EXTRA LIGHT.
4 to 6 feet per lb.	6 to 8 feet per lb.	10 to 12 feet per lb.
By Mail boxed $0.90	$0.95	$1.00

BY FREIGHT OR EXPRESS.

	HEAVY.	MEDIUM.	EXTRA LIGHT.
1 to 10 lbs.	0.45	0.55	0.65
10 " 25 "	0.43	0.53	0.63
25 " 50 "	0.41	0.51	0.61
50 " 100 "	0.40	0.50	0.60

Subject to fluctuation in price of wax.

Parties wishing to purchase large quantities, send for wholesale prices.

WM. TAYLOR,

SINCLAIR, MORGAN CO., ILL.

DADANT'S FOUNDATION.

UNSOLICITED TESTIMONIALS.

Nothing is better......Your foundation suits exactly.—C. F. Muth, Cincinnati. O., Oct. 23, 1882.

The finest and brightest I ever saw.—Green R. Shirer, Adamsville, O., March, 1882.

Thanks for promptness and the splendid article sent.—C. McRay, Canon City, Colo., June 6, 1882.

Fully equal to sample; thanks for your promptness.—D. S. Kalley, Mansfield, Ind., June 14, 1882.

Best I have used, no breaking down, bees take it readily.—W. B. Spence, Sidney, O., Aug. 31, 1882.

I like it better than any offered by dealers.—C. H. Lake, Baltimore, Md., Jan. 24, 1882.

It is the nicest I have used.—D. Keyes, Louisville, Ky., June 20, 1882.

It is the best I ever saw.—Geo. Wustum, Racine, Wis., July 19, 1882.

Willing to pay 2 cents more per pound than for any I have seen. F. Wilcox, Mauston, Wis., March 23, 1882.

The most perfect article that I have seen.—G. W. Demaree, Christiansburg, Ky.

Very well satisfied, bees worked on it fine.—Wm. Bloom, Avon, Mo.

Have used about 75 pounds from......but I prefer yours.—W. Ballantine, Sago, Muskingum Co., O.

Your foundation is the best—J. W. Porter, Charlottesville, Va., March 25, 1882.

The nicest I ever received.—H. W. Funk, Bloomington, Ill.

Your foundation beats them all. Bees draw it out faster.—Jos. Crowden, Remington, Ind.

Ahead of any foundation maker in the world.—G. M. Doolittle, Borodino, N. Y.

I concluded to send to you, even if express is higher.—D. P. Norton, Council Grove, Kans.

I never saw any nicer. G. Tisdale, Westfield, N. Y., March 31, 1882.

Better than any I have ever had.—J. B. Mason, Mechanics' Falls, Me.

We are moulding from a new set of machinery, made expressly for us. But, friends, remember that, to fill all the orders, **WE NEED WAX,** and if you have some, please write us. We pay a high price for it.

Send for our retail or wholesale circular, with samples free. We sell, also, Colonies, Queens and Supplies.

CHAS. DADANT & SON,

HAMILTON, HANCOCK CO., ILL.

CLETHRA ALNIFOLIA OR WHITE ALDER.

This shrub has attracted much notice as a forage for the honey bee, is highly recommended by eminent bee-keepers, blooms from July 1st to Sept.; the honey from it is nearly white, thick and of fine flavor, grows in any soil and is highly ornamental.

Send for our Illustrated Catalogue with full description of Clethra, prices and description of General Nursery Stock, grown on our own grounds; Souhegan Blk. Cap Raspberry, Shrubs, Shade trees, Evergreens and Herbaceous Plants specialties.

JACOB W. MANNING,

Proprietor Reading Nursery,

READING, MASS.

JOHN D. KNOX & CO.,

BANKERS AND LOAN AGENTS,

No. 202 Kansas Avenue, - Topeka, Kansas.

A GENERAL BANKING BUSINESS TRANSACTED.

EXCHANGE BOUGHT AND SOLD.

Municipal Bonds Bought and Sold. Special Attention given to Collections, and Remitted for on day of payment.

INTEREST PAID ON TIME DEPOSITS.

We issue certificates of deposit, bearing the following rates:

6 months, . . . 4 per cent. per annum.

12 months, . . . 5 per cent. per annum.

MONEY LOANED FOR INVESTORS.

On First Mortgage, at 6 to 8 per cent., on real estate worth at least three times the amount of the loan. Interest payable semi-annually. Interest and principal remitted free of exchange.

Correspondence solicited from guardians of trust funds, banks and bankers, savings associations, and others desiring safe investments with liberal rates of interest.

All letters of inquiry answered promptly, and references given. Any person having money to invest for himself or others, whether the amount is great or small, is invited to correspond with us. Safe investments, with a profitable and regular income to the lender, can be secured through us.

We should be pleased to correspond with parties contemplating real-estate investments or immigration to the State, and trust our readers will feel no hesitation in addressing us for any regular information. Having been in Kansas for sixteen years, and travelled extensively throughout the State, we have had opportunities of becoming well informed as to the comparative advantages of its various sections.

Send for a free copy of *Knox's Investor's Guide.*

JOHN D. KNOX & CO., Topeka, Kan.

Persons visiting Kansas and Topeka are cordially invited to call on us.

BARNES'
PATENT FOOT & STEAM POWER MACHINERY,

Complete outfits for Actual Workshop Business: Lathes for Wood
or Metal, Circular Saws, Scroll Saws, Formers,
Mortisers, Tenoners, etc. etc.

MACHINES ON TRIAL IF DESIRED.

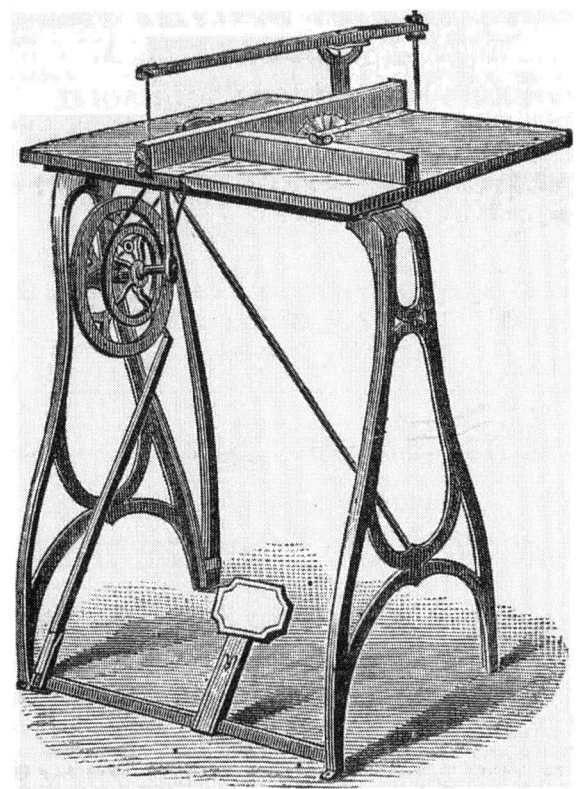

The "Combined Machine" shown by this cut is especially adapted to the
uses of Bee-keepers, and it serves their purposes completely. We give an
extract from a letter that is descriptive of its use, and it is a fair sample of
many such sent us by those using this machine.

Mr. J. I. Parent of Charlton, N. Y., says: "We cut out with one of your
"Combined Machines" last winter 50 chaff hives with 7 in. cap. 100 honey
racks, 500 brood frames, 2000 honey boxes and a great deal of other work.
We have double the amount to do this season and expect to do it with this
machine. It will do all you represent every time."

Send for catalogue and full description. Address

W. F. & JOHN BARNES,
No. 2216 Main Street, Rockford, Ill.

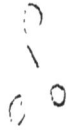

www.ingramcontent.com/pod-product-compliance
Ingram Content Group UK Ltd.
Pitfield, Milton Keynes, MK11 3LW, UK
UKHW021101110325

455992UK00010B/421